Streamlining Organizational Processes Through AI, IoT, Blockchain, and Virtual Environments

Marisol Hernández Hernández
Universidad Autónoma del Estado de México, Mexico

IGI Global
PUBLISHER of TIMELY KNOWLEDGE

A volume in the Advances in Business Information Systems and Analytics (ABISA) Book Series

Published in the United States of America by
IGI Global
Business Science Reference (an imprint of IGI Global)
701 E. Chocolate Avenue
Hershey PA, USA 17033
Tel: 717-533-8845
Fax: 717-533-8661
E-mail: cust@igi-global.com
Web site: http://www.igi-global.com

Library of Congress Cataloging-in-Publication Data

Names: Hernandez Hernandez, Marisol, 1971- editor.
Title: Streamlining organizational processes through AI, IoT, blockchain, and virtual environments / edited by: Marisol Hernández Hernández.
Description: Hershey, PA : Business Science Reference, [2023] | Includes bibliographical references and index. | Summary: "The objective of this book will be to propose optimization strategies in organizations, through technologies such as AI, IoT, Blockchain, and Virtual Environments. It will be written for professionals who wish to improve their knowledge regarding the adaptation of this type of technology in their production processes and/or services in public or private organizations, or who need to implement some ideas in their academic fields or their work areas"-- Provided by publisher.
Identifiers: LCCN 2023014952 (print) | LCCN 2023014953 (ebook) | ISBN 9781668486399 (hardcover) | ISBN 9781668486405 (paperback) | ISBN 9781668486412 (ebook)
Subjects: LCSH: Electronic data processing. | Organizational effectiveness--Technological innovations. | Operations research--Data processing.
Classification: LCC T57.5 .S784 2023 (print) | LCC T57.5 (ebook) | DDC 004--dc23/eng/20230425
LC record available at https://lccn.loc.gov/2023014952
LC ebook record available at https://lccn.loc.gov/2023014953

This book is published in the IGI Global book series Advances in Business Information Systems and Analytics (ABISA) (ISSN: 2327-3275; eISSN: 2327-3283)

British Cataloguing in Publication Data
A Cataloguing in Publication record for this book is available from the British Library.

All work contributed to this book is new, previously-unpublished material.
The views expressed in this book are those of the authors, but not necessarily of the publisher.

For electronic access to this publication, please contact: eresources@igi-global.com.

Advances in Business Information Systems and Analytics (ABISA) Book Series

ISSN:2327-3275
EISSN:2327-3283

Editor-in-Chief: Madjid Tavana La Salle University, USA

MISSION

The successful development and management of information systems and business analytics is crucial to the success of an organization. New technological developments and methods for data analysis have allowed organizations to not only improve their processes and allow for greater productivity, but have also provided businesses with a venue through which to cut costs, plan for the future, and maintain competitive advantage in the information age.

The **Advances in Business Information Systems and Analytics (ABISA) Book Series** aims to present diverse and timely research in the development, deployment, and management of business information systems and business analytics for continued organizational development and improved business value.

COVERAGE

- Business Models
- Business Decision Making
- Strategic Information Systems
- Statistics
- Forecasting
- Business Process Management
- Algorithms
- Decision Support Systems
- Data Analytics
- Legal information systems

IGI Global is currently accepting manuscripts for publication within this series. To submit a proposal for a volume in this series, please contact our Acquisition Editors at Acquisitions@igi-global.com or visit: http://www.igi-global.com/publish/.

The Advances in Business Information Systems and Analytics (ABISA) Book Series (ISSN 2327-3275) is published by IGI Global, 701 E. Chocolate Avenue, Hershey, PA 17033-1240, USA, www.igi-global.com. This series is composed of titles available for purchase individually; each title is edited to be contextually exclusive from any other title within the series. For pricing and ordering information please visit http://www.igi-global.com/book-series/advances-business-information-systems-analytics/37155. Postmaster: Send all address changes to above address.

Titles in this Series

For a list of additional titles in this series, please visit:
www.igi-global.com/book-series/advances-business-information-systems-analytics/37155

Cases on Enhancing Business Sustainability Through Knowledge Management Systems
Meir Russ (University of Wisconsin-Green Bay, USA)
Business Science Reference • © 2023 • 300pp • H/C (ISBN: 9781668458594) • US $240.00

Data-Driven Approaches for Effective Managerial Decision Making
Anubha (Jaipuria Institute of Management, Ghaziabad, India) and Himanshu Sharma (IILM Institute for Higher Education, New Delhi, India)
Business Science Reference • © 2023 • 336pp • H/C (ISBN: 9781668475683) • US $255.00

Handbook of Research on AI and Machine Learning Applications in Customer Support and Analytics
Md Shamim Hossain (Hajee Mohammad Danesh Science and Technology University, Bangladesh) Ree Chan Ho (Taylor's University, Malaysia) and Goran Trajkovski (Western Governors University, USA)
Business Science Reference • © 2023 • 423pp • H/C (ISBN: 9781668471050) • US $325.00

Building Secure Business Models Through Blockchain Technology Tactics, Methods, Limitations, and Performance
Shweta Dewangan (ICFAI University, Raipur, India) Sapna Singh Kshatri (Shri Shankaracharya Institute of Professional Management and Technology, India) Astha Bhanot (Princess Nourah Bint Abdulrahman University, Saudi Arabia) and Mushtaq Ahmed Shah (Lovely Professional University, India)
Business Science Reference • © 2023 • 283pp • H/C (ISBN: 9781668478080) • US $250.00

Enhancing Business Communications and Collaboration Through Data Science Applications
Nuno Geada (ISCTE - University Institute of Lisbon, Portugal) and George Leal Jamil (Informações em Rede Consultoria e Treinamento, Brazil)
Business Science Reference • © 2023 • 262pp • H/C (ISBN: 9781668467862) • US $270.00

For an entire list of titles in this series, please visit:
www.igi-global.com/book-series/advances-business-information-systems-analytics/37155

IGI Global
PUBLISHER of TIMELY KNOWLEDGE

701 East Chocolate Avenue, Hershey, PA 17033, USA
Tel: 717-533-8845 x100 • Fax: 717-533-8661
E-Mail: cust@igi-global.com • www.igi-global.com

Table of Contents

Detailed Table of Contents

these indices at the global level but do not at the local level. Strategies have been devised to end crime, but few have been done to prevent it, and measures have been taken that range from analyzing the air to using social networks for it. This research was carried out to propose that crime rates be reflected in maps of any local area, which will be combined in color, as a result of the application of blurred models to determine the degree of belonging of crime rates in certain areas. The result of these maps could be used to alert the population or quickly inform the corresponding justice authorities about that area so that people are cautious.

This chapter presents a proposal for the design and simulation of an automated sanding machine for the roughing of plastic parts for the company SOLGISTIKA. S.A. DE C.V. The methodology used is fuzzy logic; the software used for the design of the machine is Solidworks for the simulation of the Proteus electronic components and for the control of the rules of the Matlab system. This research is carried out by the Division of Electromechanical Engineering of National Technology of Mexico, Higher Technological Institute of the Sierra Norte de Puebla.

The current study examines the challenges associated with smart contracts and their overall concept and offers a resolution that can effectively address all the prevailing issues. Moreover, it advocates for a shift in the current architecture employed in smart contracts and blockchain. The study identifies the problems, conducts an analysis, and presents a solution that expounds on how the proposed changes can enhance the performance of smart contracts based on a new architecture, boost their scalability, change the IoT world, and promote an eco-friendly environment by reducing pollution of natural environment and conserving energy.

In recent years, the integration of the internet of things and information technology has become the critical technological and advancement tendency to remodel the manufacturing sectors. As a novel and advanced manufacturing technology, additive manufacturing has exhibited its splendid competency to generate products with complex geometry with the help of new materials. It is a class of manufacturing technologies that enables businesses to render innovative products and services. On the other hand, blockchain is one of the elements of the continuing digital transformation that has the potential to bring on modifications to business processes and is gaining momentum recently. Although blockchain attracts a lot of attention, few studies are available on its relations to additive manufacturing. This chapter investigates the impact of blockchain on additive manufacturing.

The implementation of the IoT in the industry promotes the automation of production processes in addition to the reduction of operating costs. However, its implementation is not a trivial task because it is necessary to evaluate the integration potential, analyze the different processes where its implementation is viable, seeking the most significant impact, in addition to minimizing existing technological barriers such as providing computer equipment to the industry in addition to the interconnection of devices and machines without neglecting data security in this highly connected environment. The data collected by the different devices must be able to be visualized through indicators of interest, considering these environments as intelligent environments. This chapter seeks to generate a more assimilated theoretical notion about the development and implementation of the internet of things in the industry to help the adaptation of this type of technology in their production processes and services in public or private organizations.

Some areas of priority attention in Mexico are health and education, which open an opportunity for technology, specifically this chapter presents different projects that use augmented reality (AR), virtual reality (VR), or video games (VG), which provide an excellent contribution to education and the health for its dynamism and interactivity, these projects were developed in the faculty of basic sciences, engineering, and technology (FCBIyT). The projects presented for medicine using AR are Arachnophobia, TEAPRENDE, and SARA (food labeling), AVHI: Non-Hodkin Lymphoma uses VR, and Byoky, a video game. The projects developed for education with AR are LenSeñasMx and Industrial Revolution, the projects that uses VR are the diasporas of Tlaxcala, technological advances by James Watt and the building of a giant virtual computer with this multi-screen environment. Video games for education are Independence of Mexico and the diasporas of Tlaxcala too.

A model of educational processes called multidisciplinary gradual is presented, whose objective is to guide the teacher in the teaching-learning of STEAM educational activities using as a use case the topic of basic logic gates, which associates mathematical, algebraic, electronic, and logical knowledge, presenting the theory and strengthening it in practice. The topic of logic gates is not directly related to the middle-higher level educational plan; however, it is possible to apply and deepen it by relating the basic mathematical operations. The work describes the three stages that make up the model, achieving that students approach a problem or topic from different contexts, each discipline contributes its own knowledge and qualities, but not necessarily integrated in the common understanding of the problem, so that they develop the foundations to solve interdisciplinary and/or transdisciplinary problems. At the end, the results obtained by using this model with data from a control group and an intervention group are demonstrated with statistical analysis.

Preface

In the search for the satisfaction of the needs of the human being in all the daily aspects of his life, the processes are relevant and not only refer to the processes in industrial organizations but also those involved in education, security, government, and health, among others. Processes are a series of steps to achieve an objective that in your case could be a better product, a higher sale, a safer area, meaningful learning, or priority health, among others.

Organizations are administrative systems that emerge as a producer axis of improvements created to achieve goals or objectives. These can be in multiple forms: companies, government, educational institutions, hospitals, cultural, security, law, etc. Under this idea, it is necessary to create process support tools that are excellent service providers.

The technology has multiple presentations that help to optimize processes; optimization refers to taking the variables of some process and giving it the best treatment to generate the best possible result; Process optimization refers to the treatment of variables involved in the processes, to solve a problem in an automated way that improves the expected results and also uses fewer resources, generates easier processes and/or obtains better profits of all kinds.

Different technologies can be used to achieve process optimization, but artificial intelligence, the Internet of Things, Blockchain, and virtual environments are what this book will cover.

Artificial intelligence is essential every day, and this is because its tools make the actions that are carried out daily easier or within the reach of individuals. There are several types of tools, such as neural networks that work based on training just as humans do or fuzzy technology that treats processes through a language similar to that of humans, making processes more perfect every day.

Other technologies such as the Internet of Things that move the world with sensors converted into voice, movement, light, or air, among other sensations so to speak, give entities a plus to make them more intelligent.

The BlockChain refers to a database treated as a block of code in which information is collected, stored, and shared and in turn decentralized, actions that make that database look like a single record that generates synchronized copies, but which in turn makes copying your data, resulting in better trust, security, and transparency of shared data on a business network with more cost savings.

Virtual environments are technologies that are a source of motivation and learning, that serve as a guide to something or that simply serve as fun for people; however, this technology provides resources that can help productive, cultural, psychological, or learning processes, among others. In the processes of an organization, this technology could be an important part of the training of personnel, for example, to alleviate phobias, to facilitate meaningful learning, or to provide cultural enrichment in a specific place for tourism.

The technologies described in this book give a proposal or a perspective of what organizations could be if they implement them; Its objective is to publicize the advances of these technologies, so that organizations can take them and implement them; or, researchers can take up those ideas and expand their research; it is also to generate ideas and knowledge for students who wish to learn from them.

Companies could use the ideas and build their own technology or knowledge networks; it also serves public or private companies that implement these technologies to obtain knowledge and apply it in their academic fields and later in their work areas; all this to generate better production, service, educational or security processes.

These issues are at the forefront of a globalized world, which serves to generate equal economies and better learning, work, care, health, safety, and production environments, all under the implementation scheme of cutting-edge technology. The chapters of this book address them explicitly and innovatively, proposing new techniques or improving current ones.

Regarding artificial intelligence, which has proven to be one of the most notable applications of information technology and which is a pillar to improve the efficiency and effectiveness of service provision in companies, it shows dynamic capabilities as processes that enhance transformation and enables companies to thrive in a competitive environment; capabilities that allow companies to integrate, build, and reconfigure their internal and external competencies to adapt to changes in business environments. Taking into account optimization techniques that help companies understand human behaviors, applying the dynamism of artificial intelligence as a powerful solution to improve business performance and optimization. In addition to reviewing and discussing trends in cyber threats, cyber-attacks, and cyber security models, the authors explored some recent models in cyber security to prevent data breaches and other dangers to digital systems and privacy, such as firewalls, encryption and breach detection systems and access restrictions, crisis reaction plans and routine security assessments are some of them.

In addition to cybersecurity, the issue of social crime is also addressed, since it is one of the main problems that countries face, with consequences that harm society. Taking into account that insecurity is measured with crime rates and that some statistics show these rates at a global level but not at a local level, strategies have been devised to end crime, but few to prevent it, and measures have been taken that go from analyzing the air to using social networks for it. Regarding these strategies, the authors proposed showing the crime rates on maps available anywhere, which are combined in color, as a result of the application of fuzzy models to determine the degree of belonging to certain crimes that occurred in a certain place. The result of these maps could serve to alert the population or quickly inform the corresponding justice authorities about that area so that people are cautious. The fuzzy model was applied to a specific place with four types of crimes, which show on a map the color that indicates the belonging of a crime to a specific belonging membership and that places them in the danger zone: low, medium, and high.

In productive processes, the optimization was carried out in the proposal of an industrial sander based on the fuzzy logic theory, with which the sanding process was automated, by eliminating burrs and plastic injection points through a cutter, on the pieces to be worked on for a certain time. The main benefits of the machine design are precise to optimize the processes to increase productivity and reduce the time of the personnel in charge of the roughing process of more complex parts and with greater difficulty for external and internal rework so that it's Industrial automated design has the specific function of operating the roughing of plastic injection points, at easily adjustable speeds. Computer numerical control (CNC) was used for this design, which means that the machine is capable of measuring, automating, and controlling its own processes, and thus a task that would normally have to be carried out by several operators, it is possible that the machine carries independently. This design was adjusted with the membership function that obtains the speed and work parameters for the correct operation of the sander.

The book examines the challenges associated with smart contracts and their general concept and advocates a change in the current architecture used in smart contracts and blockchain. The study identifies the problems, carries out an analysis, and presents the solution to improve the performance of smart contracts based on a new architecture, and with it, increase its scalability and promote an ecological environment by reducing the pollution of natural resources, as well as energy saving; In addition to these benefits, lower cost, higher speed and greater functionality of smart contracts for the Internet of Things are included, although with limitations of execution in the cloud.

The researchers included in the book the theme that shows how industrial and manufacturing processes have evolved enormously in a new business era, thanks to the installations of Industry 4.0, which has accelerated the use of novel technologies such as Blockchain and that it is important and with many advantages such as security, transparency, and decentralization. The Blockchain is a technology that ensures the integrity and reliability of transaction records without a trusted third-party service provider, by having all participants in the network create, record, store, and verify transaction information together, and It has the framework to perform various applications and services based on distributed network infrastructure using security technologies including hashing, digital signing, and cryptography. It is now used by many supply chains and manufacturing industries. The Blockchain applies data structure techniques to collect and decode data from entire supply chain processes made up of customers, manufacturers, and suppliers, possibly culminating in structural alterations to entire supply chain types. This Blockchain theme was linked to additive manufacturing, proposing that together they could provide a more efficient, safe, and credible additive manufacturing process.

Concerning hyperconnection, an analysis was made of the processes related to administrative and management sciences that are developed internally in the industry and impact and interact with the complex and constant network of changes that we are experiencing in the "digital" era and "interconnected"; This is because the number of devices connected to the Internet has increased due to the Internet of Things (IoT), which is why the Industry promotes the automation of production processes with lower operating costs. The authors reviewed the aspects of IoT and how it has so organically hybridized with Industry 4.0. In addition, they showed the proper way to implement IoT in an established company without diminishing daily activities, to generate a more assimilated theoretical notion on the development and implementation of the IoT in the Industry, which helps the adaptation of this type of technology in its production processes and services in public or private organizations.

Another contribution regarding cyberspace focused on the challenge of creating clear boundaries between private individuals, companies, the military, and government organizations, since the IoT is connected to it, urgent cybersecurity is needed since cybercriminals have developed more sophisticated strategies to attack computer systems. The many aspects of cyber security problems were investigated and a solution is proposed; including the methods of prevention, discovery, and cyberattack as a preventive step.

Fog computing is a distributed computing paradigm that extends cloud computing to the edge of the network, bringing computing, storage, and networking resources closer to end users and devices, enabling faster data processing, lower latency, and improved quality of service.

Fog computing is designed to address the limitations of traditional cloud computing, which relies on centralized data centers located far from end users and devices, it is also designed to address the challenges of IoT, which involves connecting a large number of devices and sensors to the Internet; Furthermore, it can help manage the large volumes of data generated by IoT devices and enable real-time processing and analysis of this data.

This research talks about this new computing paradigm that complements cloud computing by bringing computing, connection, and memory closer to edge devices and end users, thus improving speed, adaptability, performance, cryptography, and confidentiality.

Virtual environments are very important concerning "health" and "education"; how it has been implemented for autism spectrum disorder (ASD) is written, where the child with ASD is shown some habitual behaviors, social and educational, shown with augmented reality "AR"; In addition, it details a system that explains through AR what happens in the body when there is excess sugar, sodium, calories, saturated fat, and trans-fat.

In educational terms, he comments on systems that teach Mexican Sign Language (LSM), doing it with sections of words such as colors, numbers, school supplies, and furniture, with AR; in addition, a system that shows the inventions developed in the Industrial Revolution using AR.

In terms of health prevention, Virtual Reality "VR" managed to present to the user what happens in the body of a patient with Non-Hodkin Lymphoma, it was even made with a haptic interface to give it greater realism and to know the texture of the organs that have cancer.

likewise, to show the history of the time of Viceroy Don Luis de Velasco II, when he recruited families from Tlaxcala to found towns in the north of New Spain, an area threatened by the Chichimeca War. In the same way, VR was used to take a virtual tour and see the technological advances of James Watt, installed by a small town.

The VR also shows a computer seen through multiple surround screens, using his hands to build a giant virtual computer.

"VG" Virtual Games provides a video game that presents the story of a patient with dissociative identity disorder or multiple personality disorder, showing how this disease develops; and a system that allows telling the story of important events of the Independence of Mexico and about the diasporas of Tlaxcala, using this technology.

STEAM education is a paradigm to develop skills that depend on the Interdisciplinary, Transdisciplinary, and 3) Multidisciplinary approach. The STEAM approach, which combines science, technology, engineering, art, and mathematics, has become a more effective way of teaching to achieve this goal. The inclusion of technology in education is essential to prepare students for a digital world; however, it is also important to use it efficiently, effectively, and responsibly.

A combined "techno-pedagogical" strategy was created aimed at three or more disciplines of STEAM education, where it is seen that teaching-learning with activities of this nature forms skills in students so that they develop critical, creative, and analytical thinking to solve problems and give them the ability to innovate, giving creative solutions to complex problems; all focused on solving a problem.

Marisol Hernández Hernández
Universidad Autónoma del Estado de México, Mexico

Chapter 1
Artificial Intelligence, Dynamic Capabilities, and Business Optimization:
Impact on Firm Performance

Victor Kombo
Atılım University, Turkey

Caner Asbaş
Atılım University, Turkey

Şule Erdem Tuzlukaya
Atılım University, Turkey

ABSTRACT

Artificial intelligence is a recent technological paradigm that mimics human thinking and evaluating ability. It has the potential to transfer human intelligence to informatics systems in business management and transform various business aspects such as information acquisition, product creation, and service utilization. It may also help businesses mitigate threats and capitalize on opportunities. However, some organizations have rushed into its adoption despite their limited operational skills and its high costs. In this context, dynamic capabilities are crucial for a company's growth, and artificial intelligence can transform the process. This chapter reviewed and analyzed data in the literature to reveal the performance enhancement and productivity improvement led by artificial intelligence in a way that utilizes capabilities, competencies, and resources of firms.

DOI: 10.4018/978-1-6684-8639-9.ch001

INTRODUCTION

Artificial intelligence has proven to be among the most remarkable applications of information technology which has undergone incomparable growth in the past decades (Blanchet, 2016; Brynjolfsson, Rock & Syverson, 2017; Lee, Davari, Singh, & Pandhare, 2018; Rao, Gaddam, Kurni & Saritha, 2021; Wiljer & Hakim, 2019). This technology is being adopted as a pillar to enhance efficiency and effectiveness of service delivery among firms (Sanchez-Medina & C-Sanchez, 2020). Artificial intelligence is considered to have immense potential to facilitating human intelligence and revolutionizing the acquisition of information, product creation, interaction, and utilization of products and services (Rao et al., 2021).

Dynamic capabilities are processes that enhance transformation and enable companies to thrive in a competitive environment over time (Hercheui & Ranjith, 2020; Warner & Wäger, 2019). These capabilities allow firms to integrate, build, and reconfigure their internal and external competencies to adapt to changes in business environments. In today's fast-paced market, businesses must transform to remain sustainable (Nolan, 2020). Optimization techniques help firms understand human behaviors, applying artificial intelligence dynamism as a powerful solution for improving business performance and optimization (Okul, Aksu & Orman, 2019). Artificial intelligence is a technology built on two major elements - adaptability and autonomism (Blos, da Silva & Wee, 2018). According to McCracken (2017), artificial intelligence is a scientific technology that imitates human behavior, enabling machines to think, sense, reason, and solve problems like humans. Wirth (2018) explains that artificial intelligence effectively acquires knowledge through programmed data and enhances problem-solving through machine learning, resulting in increased efficiency and reduced errors (La Porte, 2018).

MAIN FOCUS OF THE CHAPTER

Artificial Intelligence, Dynamic Capabilities, and Business Optimization

The digital age we live in, which changes both the behaviors, perspectives and perceptions of people, has many profound and important influences (Asbaş & Erdem Tuzlukaya, 2021). With the effects of this era, the ways and methods in which economic and socioeconomic relations are established, maintained and executed have also been digitized at a high rate (Afonasova et al., 2019; Karpunina et al., 2019; Mehta et al., 2020). As one of the dimensions of this digitalization and with the contributions of Industry 4.0, which has become the dominant paradigm in the industry, the adoption of artificial intelligence has begun and continues to be experienced in companies.

The primary objective of artificial intelligence adoption in companies is to improve performance by increasing production efficiency and flexibility over time (Chen, Esperança & Wang, 2022). In today's dynamic business environment, firms must focus on sustaining their competitive edge against rivals by implementing strategic approaches (Ho, Gan, Jin & Le, 2022). Artificial intelligence can be leveraged through business strategies and capabilities to mitigate potential threats and capitalize on available opportunities, ensuring greater opportunities and reduced threats for companies (Lee et al., 2018).

The rapid evolution of technology has compelled organizations to adapt and reconfigure themselves to remain relevant in the dynamic business environment (Venkatram & Geetha, 2017). To sustain themselves and stay competitive, businesses need to develop dynamic capabilities that enable them to think broadly and adapt to change (Akhtar, Khan, Tarba & Jayawickrama, 2017). In the context of artificial intelligence, dynamic capabilities are becoming increasingly important as firms seek to integrate and optimize artificial intelligence technology to improve business performance.

Artificial Intelligence

Artificial intelligence can be defined technological paradigm which tends to imitate behaviors and operations of human brain. Artificial intelligence-based technologies are designed to be able to think, sense, reason and act like human demonstrating the capability of solving problems. Artificial intelligence approaches and methods are derived from the combination of fundamental mathematical techniques and higher technological and scientific methods including computation techniques, numerical methods and mathematical – geometric modelling.

The most important feature that distinguishes artificial intelligence from other paradigms and approaches is its flexible and adaptive structure, which allows it to learn on its own, based on feedback loops where errors and deviations from reality or correctness are gradually adjusted, corrected and verified through a defined learning process. Both this ability to be taught and learning process is called machine learning. Machine learning is used to name algorithms, technologies and competency in artificial intelligence paradigms, also.

Many technologies and applications that are closely related to artificial intelligence – machine learning paradigm are in use in both literate and real sector. Natural language processing can be given as the first of these. The aim of natural language processing applications, which have intense and strong connections with grammar, syntax, and many other verbal analysis and research methods, is to ensure that information communication technologies and other tools that can run software can make sense of written texts as the human brain understands and even produce necessary sensible and applicable answers. Many natural language processing-based applications are integrated into human life and organizational processes today, from tools that detect spelling and spelling errors in Microsoft Office® software, to translation tools such as Google translate, and to chatbots such as ChatGPT even with the ability to write very advanced programs and the creativity of designing articles and stories.

Another technology related to artificial intelligence is pattern recognition. Pattern recognition is the general name of the activities and techniques applied for the detection, classification and interpretation of patterns. In this context, pattern can be defined as special data or graph particles that carry or point to a certain meaning in data sets and network models (Aggarwal & Subbian, 2014). It finds its place in the industry with many applications such as pattern recognition, image processing, product classification and faulty production detection.

The use of pattern recognition opportunities offered by artificial intelligence together with image processing applications and in a coordinated manner provides the infrastructure of machine vision technologies. Machine vision can be defined as image processing-based automatic detection and analysis technologies and methods for applications such as automatic inspection and detection, process and quality control and robot management in industry. Through machine vision, many control and management activities that were previously performed by humans are gradually abandoned in favor of the detection and interpretation ability of computers and the detection systems of machines.

With the ability to make reasonable and logical decisions that artificial intelligence has, they find a place for themselves in the pattern recognition transportation sector with autonomous vehicles. Autonomous vehicles, which can both achieve to drive themselves and make necessary warnings and corrections during the human driver's driving process, have begun to deeply affect both vehicle production and design processes as well as the other sectors such as freight, cargo and human transportation where these vehicles are used, and they continue to increase their influence.

Artificial intelligence technologies make also many of the most important contributions to organizational structure and processes in the fields of information and data management. One of these contributions is big data and data mining applications presented by artificial intelligence. These applications have not only been able to automate the information and data management processes that previously depended on human research and evaluation skills, but also render information and data in large quantity, amount and details that cannot be achieved by humans, can be processed, applied and interpreted.

Missing data and information loss, which are considered to be almost impossible to avoid in real world data (Huang, Jedynak & Bader, 2007) and may cause vital problems (Burgess, Adar & Cafarella, 2016), can be experienced in different ways in various sectors such as health (Wang et al., 2018), informatics (Lopez-Martin, Carro, Sanchez-Esguevillas & Lloret, 2017) and production (Lai, Wu, Zhang, Lu & Zhong, 2019; Mirshahvalad, Lindholm, Derlén & Rosvall, 2012). Although in some cases the problem of missing data is tried to be solved by simply ignoring it (Troyanskaya et al., 2001), in many cases this solution also produces its own problems. However, with the application of artificial intelligence with different approaches in data and information management, the problem of incomplete data and information loss can be overcome and the task of information retrieval can be fulfilled.

As it is performed in information retrieval, the relations and correlations between the data in the datasets and the elements in the network can be predicted if they have not occurred yet, or they can be completed if they have realized but not been detected yet, with the link prediction methods (Clauset, Moore & Newman, 2008; Dunlavy, Kolda & Acar, 2011; Liben-Nowell & Kleinberg, 2007; Wang, Xu, Wu & Zhou, 2014) enabled by artificial intelligence methods. Link prediction applications are used in many fields and sectors such as business partner planning (Mori, Kajikawa, Kashima & Sakata, 2012), e-commerce (Liu & Kou, 2007) and telecommunication (Huang & Lin, 2009) to provide insights and develop reasonable forecasts.

On the other side, the operation of cyber-technologies and artificial intelligence on systems directly related to each other has brought transitions and combinations between these two technological families. The first of these associations is the application of artificial intelligence and machine learning methods and techniques in the field of cybersecurity and the development of some specialized cyber-defense tools in this way (Asbaş & Erdem Tuzlukaya, 2022). Artificial neural network modeling (Krundyshev & Kalinin, 2020), pattern recognition with support vector machine method (Sun, Sebastian Cardenas, Hahn & Liu, 2021) and anomaly detection processes with convolutional neural network (Alabadi & Çelik, 2020) can be listed as examples of these specialized tools. These innovative cybersecurity technologies, which are of great importance for the strategic cybersecurity management of companies, create important sources of competitive advantage by ensuring the protection and maintenance of cyber assets, ensuring that the internal functioning of the organization is carried out in a healthy manner and creating a positive image among stakeholders (Asbaş & Erdem Tuzlukaya, 2022).

The second area where artificial intelligence and cybersecurity come together is cyberwarfare precautions and strategies, which are also directly originated from and related to cyber-defense tools and techniques (Asbaş & Erdem Tuzlukaya, 2023). The fact that organizations that are critical for countries in particular are a direct target for cyberwarfare not only poses a threat to the security of these organizations, but also can irreparably damage their competitive advantages. These organizations, some examples of which can be given as health institutions, members of the banking system and energy production and distribution companies, will provide them with significant advantages in their sectors, if they have developed automated and competent measures and strategies against cyberwarfare and its possible effects and harms in an integrated manner with artificial intelligence and machine learning

Artificial intelligence methods have an intensive strength in characterizing and detecting patterns with respect to classical and traditional methods (Asjad, Alam & Hasan, 2018; Nasir, Summerfield, Dağ & Öztekin, 2020; Wang, Yu & Chan, 2012). They have significantly effective superiorities including competence in examination of systems with very high complexities, to detect, identify and categorize pattern structures, in determining, developing and resolution of detailed mathematical relationships, by decreasing their errors with an adaptive and flexible way exploiting feedback loops based on the machine learning techniques. With these superiorities, artificial intelligence has been extensively applied across various industries for different purposes through different technological components such as autonomous robots, text processing, decision-making, image processing and recognition, voice processing, as well as vehicle automation (Devang, Chintan, Gunjan, & Krupa, 2019; Wirth, 2018). According to Hollebeek, Jaakkola and Alexander (2018), there are various tools utilized in the application of artificial intelligence, and one of the most significant tools is the use of logical deductions and search. The evolution of artificial intelligence has appeared in three stages, namely, the development of the Internet and the analysis of big data, followed by the beginning of machine learning and finally the adoption and usage of artificial intelligence in daily life (Constantinides, Henfridsson & Parker, 2018).

Artificial intelligence systems are constantly evolving, with new algorithms and technologies being developed regularly. This means that firms must possess dynamic capabilities to be able to adapt and integrate these new technologies into their existing business models. Dynamic capabilities can also help firms to identify new opportunities for innovation and growth in the artificial intelligence environment.

For example, a firm that uses artificial intelligence-powered chatbots for customer support may identify new opportunities for using artificial intelligence in other areas of their business, such as sales or marketing. By possessing dynamic capabilities, the firm can adapt and integrate these new artificial intelligence applications to improve business performance and gain a competitive advantage.

In addition, dynamic capabilities can help firms to address the ethical and social implications of artificial intelligence. As artificial intelligence systems become more complex and powerful, there is a need for firms to develop and implement responsible artificial intelligence strategies that consider the ethical implications of artificial intelligence, such as privacy, bias, and accountability (Floridi et al., 2018). By possessing dynamic capabilities, firms can adapt and integrate these ethical considerations into their artificial intelligence strategies and develop responsible artificial intelligence practices.

In business, artificial intelligence has been used for a variety of purposes, including customer service, fraud detection, risk management, supply chain optimization, and personalized marketing. artificial intelligence has various applications in businesses such as customer support via chabot, risk management in financial transactions, and predictive maintenance in manufacturing to reduce downtime and improve productivity (Kshetri, 2018). One of the key advantages of artificial intelligence is its ability to analyze vast amounts of data in real-time, providing businesses with valuable insights and helping them make informed decisions. For example, retailers can use artificial intelligence to analyze customer data and provide personalized recommendations, while healthcare providers can use artificial intelligence to analyze patient data and provide better diagnosis and treatment plans (Li et al., 2019).

Manufacturing businesses have been using artificial intelligence to improve productivity and reduce downtime. For example, General Electric (GE) uses machine learning algorithms to optimize the performance of their wind turbines, resulting in a 20% increase in efficiency. Similarly, Stellantis N. V. operating in automotive sector integrates artificial intelligence-based methods into their vehicles to increase the quality, comfort and safety, and to reduce errors originated from humans (Stellantis, 2023).

In the service sector, artificial intelligence-powered chatbots have been widely adopted by businesses to provide 24/7 customer support. Bank of America uses Erica, an artificial intelligence -powered chatbot, to provide customers with financial advice and assistance. Another example is the use of artificial intelligence in the healthcare industry, where artificial intelligence-powered virtual assistants are used to provide patients with medical advice and assistance (Xiong et al., 2020).

However, the development and implementation of artificial intelligence in businesses are not without challenges. One of the most significant challenges is a scarcity of skilled professionals with the knowledge and expertise needed to develop and implement artificial intelligence-based solutions. This has led to a growing demand for skilled artificial intelligence professionals, with companies competing to attract and retain top talent in this field (Nguyen et al., 2020). The ethical implications of artificial intelligence, such as privacy, bias, and accountability, are a major challenge. As artificial intelligence becomes more powerful, regulations and standards are needed to ensure responsible development and use (Floridi et al., 2018).

Dynamic Capabilities

Dynamic capabilities refer to a firm's ability to adapt and respond to changing business environments through the use of resources and capabilities. The concept of dynamic capabilities was first introduced by Teece, Pisano, and Shuen (1997) and has since been widely studied in the field of strategic management. Teece (2017) defines dynamic capabilities as the capacity of a company to effectively combine, develop, and adjust its internal and external competencies in response to rapidly changing environments. Firms possessing dynamic capabilities have the ability to proficiently handle and adjust to modifications occurring in their industry, such as shifts in customer preferences, emerging technological innovations, and alterations in regulatory requirements. Studies have shown that dynamic capabilities are positively related to firm performance. For example, Hong and Lee (2020) found that firms with stronger dynamic capabilities were more likely to achieve higher levels of innovation and growth. Similarly, Damanpour and Aravind (2012) found that firms with strong dynamic capabilities were better able to respond to external pressures and achieve better financial performance. Dynamic capabilities are an important concept in strategic management that refers to a firm's ability to adapt and respond to changes in the business environment. Firms with strong dynamic capabilities are better able to achieve higher levels of performance and competitive advantage.

By implementing artificial intelligence technologies, organizations can gain valuable insights into their behaviors, capacities, processes, and skills, which can help them differentiate themselves from their competitors (Wohlgemuth, 2021). As a result, many businesses are focusing on developing new capabilities through the adoption of new technologies available in the market.

The relationship between artificial intelligence and dynamic capabilities is complex and evolving (Soh & Sia, 2019). On the one hand, artificial intelligence can help companies develop dynamic capabilities by allowing them to collect and process large amounts of data, allowing them to identify trends, understand customer needs, and make informed decisions. artificial intelligence can also assist businesses in automating specific tasks, freeing up resources for more strategic activities. Dynamic capabilities, on the other hand, can help businesses develop and deploy artificial intelligence more effectively (Teece, Peteraf & Leih, 2016). Companies with strong dynamic capabilities, for example, may be better positioned to identify and respond to market changes that necessitate the use of artificial intelligence, such as emerging technologies or changes in customer preferences. Dynamic capabilities can also assist businesses in developing the internal competencies required to effectively develop and market their products (Bughin, et al., 2018).

Overall, artificial intelligence and dynamic capabilities have a mutually reinforcing relationship. Companies with strong dynamic capabilities are better positioned to develop and implement artificial intelligence solutions that can aid in market adaptation and performance improvement (Parida & Madsen, 2020). Simultaneously, artificial intelligence can assist businesses in developing and improving their dynamic capabilities by allowing them to collect and process large amounts of data, automate certain tasks, and make more informed decisions.

Business Optimization

Business optimization refers to the process of improving efficiency, reducing costs, and maximizing profits under the certain conditions and constraints by using various strategies and techniques. These strategies could include but not limited to improving supply chain management, implementing new technological trend and paradigms in operations management, and analyzing business data to make informed and well-designed decisions. The main goal of business optimization is to create a new competitive advantage or raise the existing ones, and improve the overall performance of the organization under the constraints that are encountered (Lee, et al., 2018). Optimization activities aim to improve firm performance through some specific tools such as effective communication, proper documentation, employee training, and hiring quality personnel. Adopting and implementing artificial intelligence technologies into business optimization activities can significantly reduce operational and production costs across industries (Webber, Detjen, MacLean & Thomas, 2019), and enable firms to automate tasks, forecast disruptions and control operations end-to-end (Brynjolfsson et al., 2017) through the competence of artificial intelligence methods in formulating and solving of highly complex relations and mathematical equations, as well as their adaptive and flexible structures in terms of evaluating and decision making. However, in order to be able to realize the benefits and advantages of artificial intelligence, firms, at first, should utilize their resources effectively and adapt to new technological advancements over time. Several researchers have explored the concept of business optimization in their studies. For example, Lee et al. (2018) discuss how the adoption of artificial intelligence in supply chain management can lead to business optimization. They argue that artificial intelligence can be used to analyze vast amounts of data, optimize production processes, and in this way, reduce costs.

Another example of business optimization is explored by Kshetri (2018), who discusses how artificial intelligence can be used for predictive maintenance in manufacturing. By analyzing data obtained from sensors, artificial intelligence systems can predict when equipment is likely to fail and schedule maintenance before it becomes a problem. This reduces downtime, improves efficiency, and ultimately leads to cost savings.

Despite these advantages, some organizations still face impediments in leveraging artificial intelligence due to limited operational skills and high costs of procuring and maintaining the system (Anute, Paliwal, Patel & Kandale, 2021; Warner & Wäger, 2019). Therefore, some companies may hesitate in adopting artificial intelligence due to the challenges it poses. However, business optimization is a critical component of modern business strategy, and the adoption of artificial intelligence and other technologies is becoming increasingly important for achieving predefined goals and objectives.

Triple Impact on Firm Performance

The use of artificial intelligence has been a topic of interest in recent research, with varying empirical evidence emerging on its impact on firms. Tabassum and Ghosh (2018) conducted a study to examine the effects of artificial intelligence on firm productivity, finding that the introduction of artificial intelligence systems improved employee engagement and involvement, reduced turnover, and increased productivity. The study also found that artificial intelligence significantly reduced management time and efforts, as well as increased transparency and honesty in the workplace. However, the study focused solely on the aspect of production as a dependent variable, unlike the current research that emphasizes the use of dynamic capabilities. In contrast, Hercheui and Ranjith (2020) explored the impact of artificial intelligence on the dynamic capabilities of firms through seizing, reconfiguration, and sensing. Their qualitative research involved interviews with thirteen participants knowledgeable in artificial intelligence, and the findings revealed that artificial intelligence significantly impacted the dynamism of firms by enabling them to optimize their assets, understand patterns, and identify opportunities and threats. Warner and Wäger (2019) conducted a qualitative study to determine how digital transformation could be used to build dynamic capacities within firms. Their research indicated that digital transformation is a continuous process facilitated by emerging technologies, leading to the renewal of strategies and approaches to business culture, business models, and collaboration. Taken together, these studies demonstrate the potential of artificial intelligence to enhance firm performance through increased productivity, improved employee engagement, and the development of dynamic capabilities. However, there are still challenges to overcome, including limited operational skills and high costs associated with procuring and maintaining artificial intelligence systems. Further research is needed to fully understand the benefits and limitations of artificial intelligence in business operations and to develop effective strategies for its adoption and integration.

Several studies have investigated the impact of artificial intelligence on firm performance. Mikalef, Fjørtoft and Torvatn (2019) developed a theoretical framework for artificial intelligence capabilities and found that the components of artificial intelligence capabilities match well with the resource-based view. Agarwall, Das and Swain (2022) collected secondary data from firms in the Indian telecommunications, manufacturing, and information technology sectors from 2004 to 2018 and found that artificial intelligence significantly influenced operating costs and profits. Wamba-Taguimdje, Fossu Wamba, Kala Kamdjoug and Tchatchouang Wanko (2020) reviewed 500 case studies to determine the value of artificial intelligence in business transformation projects. They found that organizations adopt artificial intelligence technology to adapt or disrupt their business environment and optimize their competitive strategies. They also found that artificial intelligence can optimize existing processes, improve industry automation, enhance information transfer, and interact with human beings. Also, Ho et al. (2022) investigated the relationship between artificial intelligence and firm performance during the COVID-19 outbreak using time-series analysis. They found that while COVID-19 had a negative influence on the stock market of artificial intelligence, artificial intelligence was deployed in firms to sustain performance in a risk environment. Huang and Hayat (2019) compared the usage of artificial intelligence in companies in Pakistan and Australia and found that both countries had adopted artificial intelligence, but Australian companies performed better in terms of efficiency and effective application of artificial intelligence technology. Overall, these studies demonstrate that artificial intelligence can significantly impact firm performance by reducing costs, improving productivity, enhancing transparency, and building dynamic capabilities. However, challenges such as high costs, limited operational skills, and impediments to adoption and usage need to be addressed for firms to fully realize the potential benefits of artificial intelligence.

Artificial intelligence has been found to influence performance of firms in various ways. The adoption of artificial intelligence in businesses has been found to help manage challenges related to technology and enable businesses to realize economic growth by enhancing their operations, such as business expansion, effectiveness, and production. Artificial intelligence has significantly impacted the dynamism of firms by understanding patterns, optimizing assets, and identifying opportunities and threats, meaning that artificial intelligence could be instrumental in carrying out a SWOT analysis of any given business. Enterprises have realized growth in their sales, total factor productivity, investment in research and development, and capital expenditure. Additionally, artificial intelligence has created employment opportunities for skilled people, facilitated market valuations, and enabled companies to cut operational costs. Furthermore, adoption of artificial intelligence is related to abnormal market and stock returns (Tabassum & Ghosh, 2018).

Artificial intelligence plays a crucial role in business operations. It allows firms to integrate, build, and reconfigure their competencies to address rapid changes in business environments (Hercheui & Ranjith, 2020), optimizing performance by neutralizing potential threats and exploiting opportunities. artificial intelligence is essential in driving sustainable performance in dynamic business environments.

To adopt and use artificial intelligence technology, firms need to consider their goals, workforce, assets, management, and stakeholders. Artificial intelligence can enhance service delivery efficiency and effectiveness, and add flexibility and efficiency to operations, ultimately increasing performance (Huang & Hayat, 2019). artificial intelligence facilitates human intelligence by revolutionizing information acquisition, product creation, interaction, and product and service utilization.

CONCLUSION

The integration of artificial intelligence technology has become increasingly relevant in the business world, as firms seek to improve their performance and competitiveness in a dynamic and rapidly changing environment. This has led to several studies that explore the impact of artificial intelligence on firm performance, optimization, and dynamic capabilities.

In conclusion, the integration of artificial intelligence technology in the business environment has been associated with improved performance and productivity. This is achieved through the adoption of dynamic capabilities, which involves the sensing, seizing, and reconfiguring of business resources to remain competitive. Business optimization acts as a mediator between artificial intelligence dynamic capabilities and performance, with employee and resource optimization being key aspects. Based on the discussion, it can be proposed that the adoption and application of artificial intelligence technology has a direct significant influence on dynamic capabilities, which in turn influences firm performance. The blending of artificial intelligence with dynamic capabilities positively influences organizational performance, and the linkage between artificial intelligence and dynamic capabilities can be mediated through business optimization. Despite challenges in artificial intelligence adoption and usage, including high installation costs and limited operational skills, progress has been made in developing African countries, where multinational corporations have invested in this technology.

Digital technologies are transforming business models across various sectors and industries. artificial intelligence and dynamic capabilities offer creative and innovative solutions that enable firms to become more competitive, attracting clients with innovative products and enhancing suitability and sustainability in the dynamic business environment. Technological resources have become more important than other business resources, such as raw materials, physical resources, human resources, and financial support. Business performance is linked to creativity and innovativeness, leading policymakers to consider incorporating digital technology into their business models.

Firms should consider artificial intelligence technology as a valuable support system that can transform their business models and enhance performance. Managers can take advantage of existing capabilities to reconfigure and add value to their operations. Strategic decisions to adopt artificial intelligence should be based on initial cost, availability of skills, and other related aspects. Adoption of artificial intelligence can improve internal processes for efficiency and effectiveness, refine business collaborations and connections to clients, suppliers, and other partners, and ultimately compete better against rivals. Researchers and practitioners should consider artificial intelligence technology a valuable support system with great potential to transform business models. Firm managers should leverage their existing capabilities to adopt new technology and reconfigure their operations to add value and improve performance. Strategic decisions to adopt artificial intelligence should be based on factors such as initial cost, availability of skills, and other aspects related to acquisition and implementation. Firms that embrace artificial intelligence and use it strategically will be better equipped to compete with their rivals in the business environment.

REFERENCES

Afonasova, M. A., Panfilova, E. E., Galichkina, M. A., & Slusarczyk, B. (2019). Digitalization in economy and innovation: The effect on social and economic processes. *Polish Journal of Management Studies*, *19*(2), 23–32. doi:10.17512/pjms.2019.19.2.02

Agarwall, H., Das, C. P., & Swain, R. K. (2022). *Does artificial intelligence influence the operational performance of companies? A study*. Atlantis Highlights in Social Sciences, Education and Humanities., doi:10.2991/ahsseh.k.220105.008

Aggarwal, C., & Subbian, K. (2014). Evolutionary network analysis. *ACM Computing Surveys*, *47*(1), 1–36. doi:10.1145/2601412

Akhtar, P., Khan, Z., Tarba, S., & Jayawickrama, U. (2017). The Internet of Things, dynamic data and information processing capabilities, and operational agility. *Technological Forecasting and Social Change*, *136*, 307–316. doi:10.1016/j.techfore.2017.04.023

Alabadi, M., & Çelik, Y. (2020). Anomaly detection for cyber-security based on convolutional neural network: A survey. *2020 International Congress on Human-Computer Interaction, Optimization and Robotic Applications (HORA)*. 10.1109/HORA49412.2020.9152899

Anute, N., Paliwal, M., Patel, M., & Kandale, N. (2021). Impact of artificial intelligence and machine learning on business operations. *Journal of Management Research and Analysis*, *8*(2), 69–74. doi:10.18231/j.jmra.2021.015

Asbaş, C., & Erdem Tuzlukaya, Ş. (2021). Dijital çağda sosyal ağlar ve sosyal girişimcilik. In D. Hıdıroğlu (Ed.), *Dijital Çağda Kurumsal Yönetişim, Sosyal Sorumluluk ve Sosyal Girişimcilik* (pp. 395–418). Artikel Akademi.

Asbaş, C., & Erdem Tuzlukaya, Ş. (2022). Cyberattack and cyberwarfare strategies for businesses. In F. Özsungur (Ed.), *Conflict Management in Digital Business: New Strategy and Approach* (pp. 303–329). Emerald. doi:10.1108/978-1-80262-773-220221027

Asbaş, C., & Erdem Tuzlukaya, Ş. (2023). Cyberwarfare: War activities in cyberspace. In F. Özsungur (Ed.), *Handbook of Research on War Policies, Strategies, and Cyber Wars* (pp. 128–145). IGI Global. doi:10.4018/978-1-6684-6741-1.ch007

Asjad, M., Alam, A., & Hasan, F. (2018). A comparative study of classifier techniques for lift index data analysis. *Benchmarking*, *25*(2), 632–641. doi:10.1108/BIJ-09-2016-0137

Blanchet, M. J. O. T. (2016). Industrie 4.0: Nouvelle donne industrielle, nouveau modèle économique. *Outre-Terre*, *46*(1), 62–85. doi:10.3917/oute1.046.0062

Blos, M. F., da Silva, R. M., & Wee, H. M. (2018). A framework for designing supply chain disruptions management considering productive systems and carrier viewpoints. *International Journal of Production Research*, *56*(15), 5045–5061. doi:10.1080/00207543.2018.1442943

Brynjolfsson, E., Rock, D., & Syverson, C. (2017). *Artificial intelligence and the modern productivity paradox: A clash of expectations and statistics* (Vol. 24001). NBER Working Paper. doi:10.3386/w24001

Bughin, J., Hazan, E., Lund, S., Dahlström, P., Wiesinger, A., & Subramaniam, A. (2018). *Skill Shift: Automation and The Future of the Workforce*. McKinsey Global Institute.

Burgess, M., Adar, E., & Cafarella, M. (2016). Link-prediction enhanced consensus clustering for complex networks. *PLoS One*, *11*(5), e0153384. Advance online publication. doi:10.1371/journal.pone.0153384 PMID:27203750

Chen, D., Esperança, J. P., & Wang, S. (2022). The impact of artificial intelligence on firm performance: An application of the resource-based view to e-commerce firms. *Frontiers in Psychology*, *13*, 1–14. doi:10.3389/fpsyg.2022.884830 PMID:35465474

Clauset, A., Moore, C., & Newman, M. E. (2008). Hierarchical structure and the prediction of missing links in networks. *Nature*, *453*(7191), 98–101. doi:10.1038/nature06830 PMID:18451861

Constantinides, P., Henfridsson, O., & Parker, G. G. (2018). Introduction – Platforms and infrastructures in the digital age. *Information Systems Research*, *29*(2), 381–400. doi:10.1287/isre.2018.0794

Damanpour, F., & Aravind, D. (2012). Managerial innovation: Conceptions, processes, and antecedents. *Management and Organization Review*, *8*(2), 423–454. doi:10.1111/j.1740-8784.2011.00233.x

Devang, V., Chintan, S., Gunjan, T., & Krupa, R. (2019). Applications of artificial intelligence in marketing. *Annals of Dunarea de Jos University of Galati Fascicle I. Economics and Applied Informatics*, *25*(1), 28–36. doi:10.35219/eai158404094

Dunlavy, D. M., Kolda, T. G., & Acar, E. (2011). Temporal link prediction using matrix and tensor factorizations. *ACM Transactions on Knowledge Discovery from Data*, *5*(2), 1–27. doi:10.1145/1921632.1921636

Floridi, L., Cowls, J., Beltrametti, M., Chatila, R., Chazerand, P., Dignum, V., Luetge, C., Madelin, R., Pagallo, U., Rossi, F., Schafer, B., Valcke, P., & Vayena, E. (2018). AI4People–An ethical framework for a good AI society: Opportunities, risks, principles, and recommendations. *Minds and Machines*, *28*(4), 689–707. doi:10.100711023-018-9482-5 PMID:30930541

Hercheui, M., & Ranjith, R. (2020). Improving organization dynamic capabilities using artificial intelligence. *The Global Journal of Business Research*, *14*(1), 87–96.

Ho, L. T., Gan, C., Jin, S., & Le, B. (2022). Artificial intelligence and firm performance: Does machine intelligence shield firms from risks? *Journal of Risk and Financial Management*, *15*(7), 302. doi:10.3390/jrfm15070302

Hollebeek, L., Jaakkola, E., & Alexander, M. (2018). Beyond the dyadic: Customer engagement in increasingly networked environments. *Journal of Service Management*, *29*(3), 330–332. doi:10.1108/JOSM-05-2018-410

Hong, J. I., & Lee, Y. G. (2020). Dynamic capabilities, environmental dynamism, and innovation performance: Evidence from the biotech industry. *Sustainability*, *12*(6), 2217.

Huang, H., Jedynak, B. M., & Bader, J. S. (2007). Where have all the interactions gone? Estimating the coverage of two-hybrid protein interaction maps. *PLoS Computational Biology*, *3*(11), e214. Advance online publication. doi:10.1371/journal.pcbi.0030214 PMID:18039026

Huang, W., & Hayat, A. (2019). Impact of artificial intelligence in enterprises HR performance in Pakistan: A comparison study with Australia. *Global Journal of Management and Business Research*, *19*(A15), 53–58.

Huang, Z., & Lin, D. K. (2009). The time-series link prediction problem with applications in communication surveillance. *INFORMS Journal on Computing*, *21*(2), 286–303. doi:10.1287/ijoc.1080.0292

Karpunina, E. K., Yurina, E. A., Cheremisina, T. N., Andryiashka, M. V., & Kuzmina, O. Y. (2019). 'Conversation Economy' as a determinant of digital development: New opportunities and latent threats. *Proceedings of the 34th International Business Information Management Association Conference*.

Krundyshev, V., & Kalinin, M. (2020). Prevention of cyber attacks in smart manufacturing applying modern neural network methods. *IOP Conference Series. Materials Science and Engineering*, *940*(1), 012011. doi:10.1088/1757-899X/940/1/012011

Kshetri, N. (2018). 1 blcokchain's roles in meeting key supply chain management objectives. *International Journal of Information Management*, *39*, 80–89. doi:10.1016/j.ijinfomgt.2017.12.005

La Porte, N. (2018). We're not giving the answers so much as asking the questions. *Fast Company*, *224*, 46–50.

Lai, X., Wu, X., Zhang, L., Lu, W., & Zhong, C. (2019). Imputations of missing values using a tracking-removed autoencoder trained with incomplete data. *Neurocomputing*, *366*, 54–65. doi:10.1016/j.neucom.2019.07.066

Lee, J., Davari, H., Singh, J., & Pandhare, V. (2018). Industrial artificial intelligence for industry 4.0-based manufacturing systems. *Manufacturing Letters*, *18*, 20–23. doi:10.1016/j.mfglet.2018.09.002

Liben-Nowell, D., & Kleinberg, J. (2007). The link-prediction problem for social networks. *Journal of the American Society for Information Science and Technology*, *58*(7), 1019–1031. doi:10.1002/asi.20591

Liu, Y., & Kou, Z. (2007). Predicting who rated what in large-scale datasets. *SIGKDD Explorations*, *9*(2), 62–65. doi:10.1145/1345448.1345462

Lopez-Martin, M., Carro, B., Sanchez-Esguevillas, A., & Lloret, J. (2017). Network traffic classifier with convolutional and recurrent neural networks for Internet of Things. *IEEE Access : Practical Innovations, Open Solutions*, *5*, 18042–18050. doi:10.1109/ACCESS.2017.2747560

McCracken, H. (2017). The great AI war of 2018. *Fast Company*, *220*, 64–73.

Mehta, R., Singh, H., Banerjee, A., Bozhuk, S., & Kozlova, N. (2020). Comparative analysis of the consequences of purchasing models transformation within the global digitalization of the economy. *IOP Conference Series. Materials Science and Engineering*, *940*(1), 012071. doi:10.1088/1757-899X/940/1/012071

Mikalef, P., Fjørtoft, S. O., & Torvatn, H. Y. (2019). Developing an artificial intelligence capability: A theoretical framework for business value. Business Information Systems Workshops: BIS 2019 International Workshops. doi:10.1007/978-3-030-36691-9_34

Mirshahvalad, A., Lindholm, J., Derlén, M., & Rosvall, M. (2012). Significant communities in large sparse networks. *PLoS One*, *7*(3), e33721. Advance online publication. doi:10.1371/journal.pone.0033721 PMID:22479433

Mori, J., Kajikawa, Y., Kashima, H., & Sakata, I. (2012). Machine learning approach for finding business partners and building reciprocal relationships. *Expert Systems with Applications*, *39*(12), 10402–10407. doi:10.1016/j.eswa.2012.01.202

Nasir, M., Summerfield, N., Dağ, A., & Öztekin, A. (2020). A service analytic approach to studying patient no-shows. *Service Business*, *14*(2), 287–313. doi:10.100711628-020-00415-8

Nguyen, T. T., Nguyen, Q. V. H., Nguyen, D. T., Yang, S., Eklund, P. W., Huynh-The, T., Nguyen, T. T., Pham, Q. V., Razzak, I., & Hsu, E. B. (2020). Artificial intelligence in the battle against coronavirus (COVID-19): A survey and future research directions. arXiv preprint arXiv:2008.07343

Nolan, A. (2020). Artificial intelligence, digital technology and advanced production. In OECD (Ed.), The Digitalisation of Science, Technology and Innovation: Key Developments and Policies (pp. 119–142). OECD Publishing.

Okul, Ş., Aksu, D., & Orman, Z. (2019). Investigation of artificial intelligence based optimization algorithms. *Journal of Istanbul Sabahattin Zaim University Natural Sciences Institute*, *1*(1), 11–16.

Parida, V., & Madsen, E. L. (2020). Artificial intelligence for dynamic capabilities: Empirical evidence from manufacturing firms. International Journal of Operations & Production Management.

Rao, T., Gaddam, A., Kurni, M., & Saritha, K. (2021). Reliance on artificial intelligence, machine learning and deep learning in the Era of Industry 4.0. In Smart Healthcare System Design: Security and Privacy Aspects (pp. 281–299). Wiley.

Sanchez-Medina, A. J., & C-Sánchez, E. (2020). Using machine learning and big data for efficient forecasting of hotel booking cancellations. *International Journal of Hospitality Management*, *89*, 102546. doi:10.1016/j.ijhm.2020.102546

Soh, P. H., & Sia, S. K. (2019). A dynamic capabilities perspective on the strategic value of artificial intelligence: Implications for organizational learning and innovation. *The Journal of Strategic Information Systems*, *28*(1), 1–12.

Stellantis. (2023). *Autonomous driving*. Stellantis. https://www.stellantis.com/en/technology/autonomous-driving

Sun, C. C., Sebastian Cardenas, D. J., Hahn, A., & Liu, C. C. (2021). Intrusion detection for cybersecurity of Smart Meters. *IEEE Transactions on Smart Grid*, *12*(1), 612–622. doi:10.1109/TSG.2020.3010230

Tabassum, F., & Ghosh, N. (2018). Future of performance management in artificial intelligence ERA. *Shanlax International Journal of Arts. Science and Humanities*, *6*(S1), 7–14. doi:10.5281/zenodo.1469447

Teece, D. J. (2017). Explicating dynamic capabilities: The nature and microfoundations of (sustainable) enterprise performance. *Strategic Management Journal*, *28*(13), 1319–1350. doi:10.1002mj.640

Teece, D. J., Peteraf, M., & Leih, S. (2016). Dynamic capabilities and organizational agility: Risk, uncertainty, and strategy in the innovation economy. *California Management Review*, *58*(4), 13–35. doi:10.1525/cmr.2016.58.4.13

Teece, D. J., Pisano, G., & Shuen, A. (1997). Dynamic capabilities and strategic management. *Strategic Management Journal*, *18*(7), 509–533. doi:10.1002/(SICI)1097-0266(199708)18:7<509::AID-SMJ882>3.0.CO;2-Z

Troyanskaya, O., Cantor, M., Sherlock, G., Brown, P., Hastie, T., Tibshirani, R., Botstein, D., & Altman, R. B. (2001). Missing value estimation methods for DNA microarrays. *Bioinformatics (Oxford, England)*, *17*(6), 520–525. doi:10.1093/bioinformatics/17.6.520 PMID:11395428

Venkatram, K., & Geetha, M. A. (2017). Review on big data analytics – Concepts, philosophy, process and applications. *Cybernetics and Information Technologies*, *17*(2), 3–27. doi:10.1515/cait-2017-0013

Wamba-Taguimdje, S. L., Fossu Wamba, S., Kala Kamdjoug, J. R., & Tchatchouang Wanko, C. E. (2020). Influence of artificial intelligence (AI) on firm performance: The business value of AI-based transformation projects. *Business Process Management Journal*, *26*(7), 1893–1924. doi:10.1108/BPMJ-10-2019-0411

Wang, P., Xu, B. W., Wu, Y. R., & Zhou, X. Y. (2014). Link prediction in social networks: The state-of-the-art. *Science China. Information Sciences*, *58*(1), 1–38. doi:10.100711432-014-5237-y

Wang, Y., Wang, L., Rastegar-Mojarad, M., Moon, S., Shen, F., Afzal, N., Liu, S., Zeng, Y., Mehrabi, S., Sohn, S., & Liu, H. (2018). Clinical Information Extraction Applications: A literature review. *Journal of Biomedical Informatics*, *77*, 34–49. doi:10.1016/j.jbi.2017.11.011 PMID:29162496

Wang, Y., Yu, C. Y., & Chan, H. H. (2012). Predicting construction cost and schedule success using artificial neural networks ensemble and support vector machines classification models. *International Journal of Project Management*, *30*(4), 470–478. doi:10.1016/j.ijproman.2011.09.002

Warner, K. S. R., & Wäger, M. (2019). Building dynamic capabilities for digital transformation: An ongoing process of strategic renewal. *Long Range Planning*, *52*(3), 326–349. doi:10.1016/j.lrp.2018.12.001

Webber, S. S., Detjen, J., MacLean, T. L., & Thomas, D. (2019). Team challenges: Is artificial intelligence the solution? *Business Horizons*, *62*(6), 741–750. doi:10.1016/j.bushor.2019.07.007

Wiljer, D., & Hakim, Z. (2019). Developing an artificial intelligence–enabled health care practice: Rewiring health care professions for better care. *Journal of Medical Imaging and Radiation Sciences*, *50*(4, Supplement 2), 8–14. doi:10.1016/j.jmir.2019.09.010 PMID:31791914

Wirth, N. (2018). Hello marketing, what can artificial intelligence help you with. *International Journal of Market Research*, *60*(5), 435–438. doi:10.1177/1470785318776841

Wohlgemuth, V. (2021). Dynamic capabilities: Celebrating the plurality of understandings of the concept. In T. Bayón, M. Eisend, J. Koch, A. Söllner, M. Vodosek, & H. T. Wagner (Eds.), *Dynamic Capabilities and Relationships: Discourses, Concepts, and Reflections* (pp. 1–28). Springer.

Xiong, Z., Wu, Q., Zhao, X., Zuo, J., Xiong, H., & Xie, B. (2020). the application of artificial intelligence in healthcare: A comprehensive review and critique. *Journal of Healthcare Engineering*, *2020*, 1–21. doi:10.1155/2020/2973671

KEY TERMS AND DEFINITIONS

Artificial Intelligence: Technological approaches and paradigm which imitate thinking behaviors of human beings. Artificial intelligence-integrated systems are able to think, sense, reason and act like humans demonstrating the capability of solving problems similar to human beings.

Dynamic Capabilities: Basic routine processes that progress transformation among companies through evolution with time making them survive well in a competitive environment.

Firm Performance: End results achieved by an organization either in financial or non-financial form. In the current research, firm performance is measured in terms of various aspects including customer satisfaction, employee turnover, increased productivity, market share, and innovation/new products.

Optimization: Process of improving efficiency, reducing costs, and maximizing profits under the certain conditions and constraints by using various strategies and techniques. It is generally centered through improvement of communication among employees and management, proper documentation, improving employees' skills through training and hiring quality manpower in various positions.

Reconfiguration: The ability of a firm optimizing its available assets appropriately and effectively.

Seizing: The aspect of a firm understanding adopting business patterns such as utilizing of a new technology by an organization which desires to adapt to the environment.

Sensing: The ability of an organization to identify available opportunities and to notice potential threats.

Chapter 2
Fuzzy Models to Show Crime Rates on Maps

Marisol Hernández Hernández
Universidad Autónoma del Estado de México, Mexico

Luis Alfonso Bonilla Cruz
Tecnológico de Estudios Superiores de Ixtapaluca, Mexico

Lizbeth Cobian Romero
Tecnológico de Estudios Superiores de Ixtapaluca, Mexico

ABSTRACT

Crime is one of the main problems facing countries with consequences that harm society. Insecurity is measured with crime rates; some statistics show these indices at the global level but do not at the local level. Strategies have been devised to end crime, but few have been done to prevent it, and measures have been taken that range from analyzing the air to using social networks for it. This research was carried out to propose that crime rates be reflected in maps of any local area, which will be combined in color, as a result of the application of blurred models to determine the degree of belonging of crime rates in certain areas. The result of these maps could be used to alert the population or quickly inform the corresponding justice authorities about that area so that people are cautious.

INTRODUCTION

The crime is any legally punishable action that seriously offends the ethical-legal order and therefore deserves a sanction; jurists conceive it as a topical and guilty unlawful act (Torres and Granadillo, 2019). There are several types of crimes, some

DOI: 10.4018/978-1-6684-8639-9.ch002

belong to the commercial, civil, labor area; however, those that affect citizenship are classified as criminal offenses, which are serious and affect the integrity of people.

Crime is one of the main problems in countries and its consequences are detrimental to society. The crime incidence shows some statistics at a global level, but they do not at a local level. Some crimes classified as serious are kidnapping, intentional homicide, robbery, rape, and drug trafficking and according to the federal penal codes of several countries, they are the ones that cause the most damage to society.

In Mexico, it seems that the crime rate has been constant, for an average of 35,236 crimes occurred from 2010 to 2021, according to data reported by Geografía INEGI (2022); However, that publication shows that although 30,535 occurred in 2010 and 30,716 in 2021, there have been periods where a drastic increase has been observed, for example in the years 2015 to 2016. Then, in 2017, that can be seen in graph 1.

The graph shows that although in recent years there has been a constant rate of crime, it has not decreased much and actions are required to stop it from increasing.

In the United States, the average crime rate in 2020 was 398.5 crimes per 100,000 inhabitants; a slight increase compared to 2019, where the figure was 380.8, according to CNN newscast, Haczek (2022). It can also be highlighted on this subject, that in the world in the year 2022, the newspaper "la República" classified the countries with the highest crime according to the image of Figure 1.

Figure 1. The crime rate in Mexico
Source INEGI (2022)

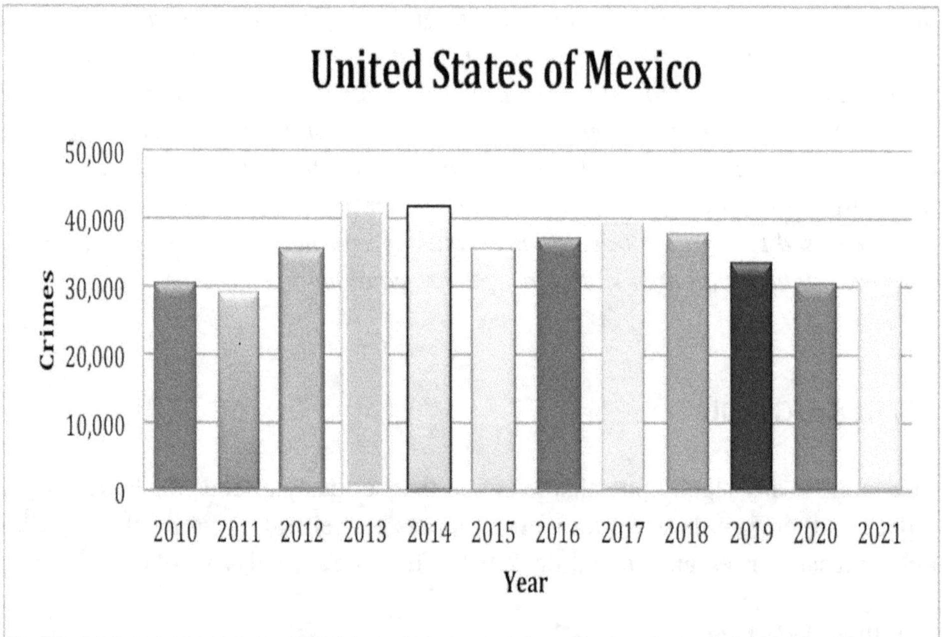

Figure 2. The crime rate in the world
Source S.A.S, E. L. R. (n.d.).

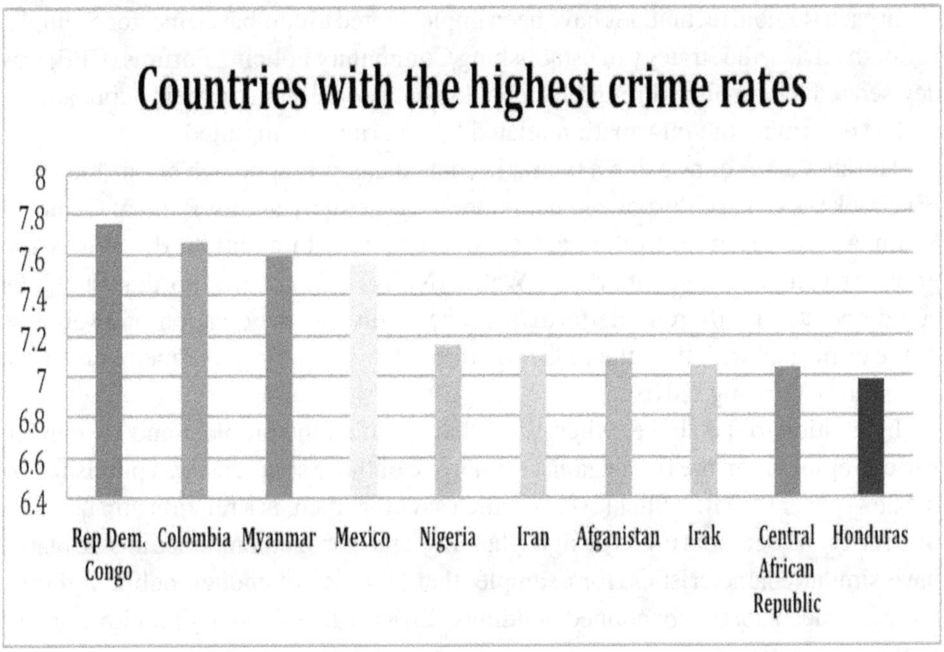

Each zone has its classifications, depending on various factors. Some are described below to give more contexts to the impact of crime:

- In the case of South Africa, which in 2015 had one of the highest crime rates in the world (Hatting, 2015).
- In Peru, in 2016, 2,435 deaths were registered in intentional criminal acts, which showed an index of 7.7 deaths per hundred thousand inhabitants, with a homicide rate of 7.7 deaths per 100,000 inhabitants (INEI, 2016).
- Latin America is the area of the world where the most malicious crimes are committed (Hernández, 2021); The factors that affect this area and that result in this assertion are organized crime, youth gangs, and little or no action by institutions dedicated to fighting crime.

In the analysis of the way in which the fight against crime is carried out, there are several contexts, such as Crime Prevention, which is combined with formal theories and policies that include the use of ICTs, aimed at crime prevention. (Palmer, Warren & Miller (2018); which is carried out through surveillance, used to monitor people through the convergence of data through discreet surveillance and through information and communication technologies.

In each Region, techniques have been implemented to combat crime, for example, in South Africa, the strategy of establishing Community Policing Forums (CPF) was devised, which monitor through Facebook what people share in their publications, to thus determine the information related to the crimes committed.

The above was done using a thematic content analysis approach of one month of Facebook posts in a CPF-policed community, particularly in East Rand of Gauteng, South Africa, resulting in the analysis of 247 posts; In addition, the WhatsApp group of your local Neighborhood Watch (NHW) was created, so that the CPFs could operate on different platforms and the result was the creation of awareness of the crime and with that, the reduction through the actions of the members of the community (Hatting, 2015).

In addition to the above, other places have studied the problem and have given various opinions on the fight against crime, one of those strategies was published by the newspaper ONU-Habitat (2022), which says that there is a relationship between insecurity and deficient urban design, planning, and management, since unsafe places have similar characteristics, for example, that there is not enough public lighting, there are vacant lots or abandoned buildings, difficult access, poorly monitored areas or that residential, work and commercial spaces are not combined.

In other research there has been a mapping of data on crime in established sociology which digitally links the theories of crime being worked on, digitally mapping various records of convicted criminals in Pakistan on social theories leading societies to commit crimes groups in their respective communities, which has led sociologists to curb crime in communities through the correct intervention of policies, incentives, and compliance Khalid, Khan, & Ifzal (2021).

Other authors have emphasized strategies for crime prevention, giving it greater importance and saying that social crime strategies must be established through the union of the police and the Criminal Justice System, to get to a certain point in meeting the crime prevention challenge (Grant, 2015).

Therefore, it is observed that there several factors help increase the crime rate, including air quality, since Bondy, Roth and Sager (2020) and Lutz et al.,(2019) investigated and found that air pollution significantly impacts crime statistics, including those crimes whose motive of being is economic.

Demir and Ozcan (2022) investigated the results of the United Nations surveys on crime trends and which was carried out based on the World Development Indicators, these data were collected from 37 countries that belonged to the OECD during the period 2003-2017; the investigation found that increases in green spaces reduce the crime rate and conversely, the same does not happen when air pollution increases.

As can be seen, various approaches have been taken to combat crime since the criminal justice system normally attacks it from the beginning, legislating on what is considered a crime and at the other extreme, which are arrests and individual prosecutions; however, Situational Crime Prevention (SCP) is rarely carried out, which can address topics ranging from the dispatch of harmful substances to general disorganization or injustice in a community (Freilich and Newman, 2017).

In this context, more strategies should be devised for the prevention and/or combat of crimes or, give more ideas to provide information to the population, in order to take care of being a victim of them. Organizations dedicated to fighting crime need to use tools that allow them to improve their activities. Prosecutor's offices collect statistics on criminal incidence on a daily basis, but processes are needed to help optimize these activities.

Concatenated with the above it is known that, the mission of the organizations that are dedicated to the prevention and combat of crime, are the municipal, State, Federal police, in their different categories. Crimes vary from place to place and from time to time; fuzzy logic would help to optimize prevention processes, helping the population to determine for themselves the conflict zones.

So too, the International Cartographic Association conceives cartography as a set of studies and scientific and technical operations that intervene in the formation and analysis of maps (Campos, 1990, cited by Franco and Valdez, 2002).

Under this tenor, cartography is a global system that serves to locate various elements. The new technology that includes the American system of navigation and location by satellites 'GPS' makes maps essential in everyday life. GPS is an element that has given way to real-time locations and that, together with other tools that show the context of those places; make your consultation an ideal place to avoid getting lost in space.

Derived from all of the above, this research proposes that a cartographic system based on fuzzy models be carried out, which locates the areas according to their crime incidence, translating it into maps that draw any local area and that are combined in color, according to the degree of belonging determined by the fuzzy model.

Besides that, the result of the application of fuzzy models will determine the sharpness in the colors of the maps, which will be decided according to their criminality incidence in low, medium, high, low-medium, and medium-high.

Other additional results would be:

- These maps would serve to alert the population or quickly inform the corresponding justice instances in that area so that people are cautious and that the justice agencies apply strategies to combat crime.
- The membership function will allow knowing "the extent" to which the crime rate of a specific crime occurs and visualize its variations in real-time.

- After carrying out the simulation of the maps based on fuzzy logic, the result obtained was to show the colored maps according to the degree of belonging of the crimes committed in the area.
- The maps are shown in blue when the crime membership is low, in orange when it is medium, and in green when it is high; Thus, when the membership of the crimes is between low and medium, the maps will be painted blue and yellow, when it is between medium and high, the map will be painted green and yellow.
- The degree of color will depend on the degree of membership of the crimes, that is, if the low membership is greater than the average, the blue color will be seen more than the orange, if the highest membership is the highest one, the map it will be painted green, in greater proportion than an orange.

In this way, the degree of delinquency can be observed more easily for the people who consult it; whether civil or police. This would make it possible to optimize the processes of prevention, detection and combating crime by the organizations that are dedicated to this activity.

Background

The fuzzy logic proposed by Lofti A. Zadeth arises with the purpose of inferring that human language is imprecise when it comes to describing phenomena that are not well defined; states that human thinking, particularly in the domains of pattern recognition, information communication, and abstraction, is imprecise (Zadeh, 1965).

Diffuse information is located in degrees of membership in one set or another, which means in linguistic terms, that not everything is true and not everything is false; For example, if the ambient temperature is hot, it would be necessary to know who is saying it, where they are saying it, or for what application it is being deduced; Because a person in a hot region will say that hot is 45° Celsius and a person who lives in a cold region could say that 10° Celsius is already hot, and both people will be right, that is why the information is inaccurate or vague.

To categorize elements into small or large with hard data, one would say drastically and according to an established criterion, what is small or what is large; however, to what extent are they classified as big or small? or against whom are they being compared?; This is based on human reasoning.

Most of the logic of human reasoning is not the classical logic of two values, or even of several values, rather it is a logic of fuzzy truths, fuzzy conjunctions or fuzzy rules of deduction" (Zadeh, 1977). This logic includes terms such as age, membership, membership, which will be described to give a greater context to the research description.

Vagueness is statements that are not true or false but must be qualified in degrees of reliability (Ramos-Soto and Pereyra-Fariña, 2017). Fuzzy logic handles the vagueness of sets according to their degree of membership; that is, classical logic would classify all people as wide or thin; in fuzzy logic, it gives them a certain degree of membership.

Under this reasoning, each person belongs to the wide set and in turn to the thin set to a greater or lesser degree, depending on what they are classifying for, and this is called the degree of belonging or membership. This element is a fundamental part of models based on fuzzy logic.

Fuzzy systems are represented with fuzzy sets, whose main element is membership, which establishes degrees of membership with values between [0,1], establishing a gradual membership and establishing that it includes hard values that are 0 and 1.

Fuzzy sets have vacuity, which means that the sets are empty, this means that their membership is 0 (Emptyness), and their complement is the amount that the membership needs to reach 1.

Fuzzy sets are also joined and intersected, following the following methods:

Union:

$$A \cup B(x) = A(x) \cup B(x) = \text{máx}\{A(x), B(x)\}$$

Intersection:

$$A \cap B(x) = A(x) \cap B(x) = \min\{A(x), B(x)\}$$

Set negation:

$$\bar{A}\,(x) = A(x) \vee B(x) = \neg 1 - A(x)$$

Membership is translated linguistically with If-Then-Else statements, for example, for a certain problem the following rules may arise:

If the temperature is >10° Celsius, it is cold, if it is between 11 and 20, it is warm and if it is greater than 20 it is hot, in such a way that the fuzzy graph is the one shown in Figure 3.

There are multiple investigations based on fuzzy logic, most located in engineering, others in other contexts such as social, political, or health; some are mentioned below:

Veerasamy and Sangeetha (2022) proposed an algorithm that uses macroblock-based searching; what it does is maintain the pixel resolution so that the quality is higher in streaming media and in increasing the maximum signal-to-noise ratio (PSNR).

Figure 3. Membership features
Source: Own

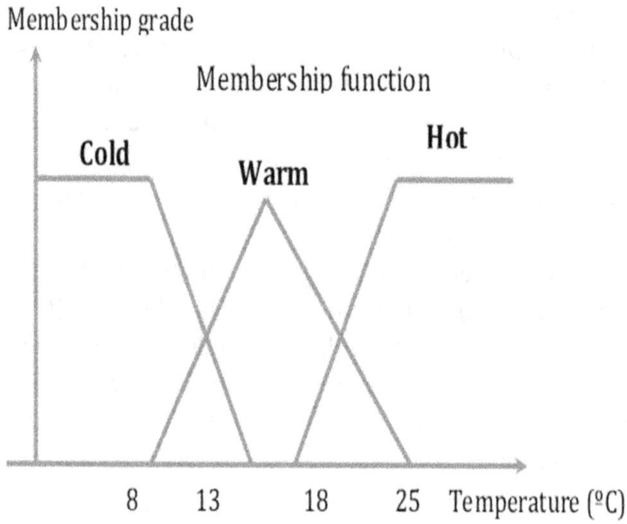

Other applications are to help medical diagnosis, such as the case of the cancer detection analysis proposed by Halder and Kumar (2019), who determined that their technique, through uncertainty, compared cancer data sets, showing superiority for prediction over other cutting-edge techniques.

Another field of application of fuzzy logic is fuzzy expert systems, such as the one carried out by (Cardona, Restrepo, & Ovalle, 2007), called Fuzzy Profile, which is a fuzzy expert system, designed to treat patients with criminal records who need to be evaluated to determine if they are fit to interact with a society whose results are given in natural language and is designed to be used by any Institution you need to know about these profiles.

Flores and Vallejo (2015), in their research on social phenomena and policy evaluation, concluded that the fuzzy logic theory can be very useful for dealing with social phenomena, due to its abstract and practical understanding, which reduces it to uncertainty and vagueness of the decisions of human thought, which redirects it to social intervention to improve the analytical vision in the evaluation of social programs.

In addition to applying logic to electronic systems, image detection, or social problems, fuzzy information systems are made. Wang, Chen, and Hu (2014), define them as systems that have relationships between the attributes of information systems and the vagueness of a set.

The fuzzy systems premise is based on Figure 1, where the basic components of type 1 fuzzy system are shown, whose characteristics must be taken into account for the realization of fuzzy systems.

Figure 4. Diffuse system
Source: Own

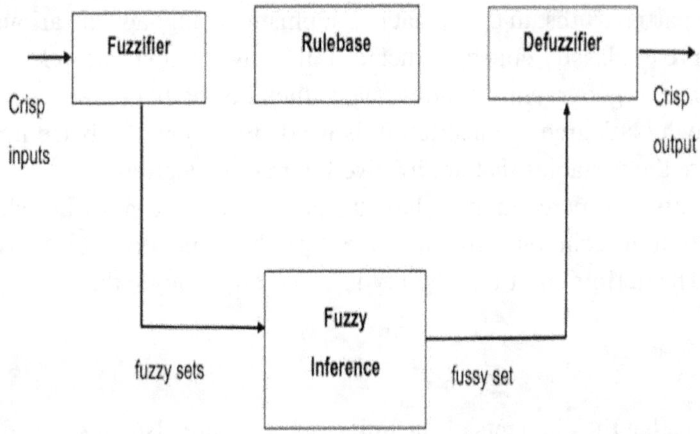

The figure has elements that guide the context of fuzzy systems, which are detailed below.

Fuzzifier

In this section, the real values are transformed linguistic variables; the input and output values are defined by the input and output variables and the degree of membership of the attributes is expressed mathematically.

Rulebase

Conditional clauses are created by the user or subject matter expert to whom the fuzzy rules are applied and can take the form:

If A then B

If A and B, then C

fuzzy inference: Defines the behavior of a system through the use of rules of type <When>, <Then> at the linguistic level. That is, it takes some data and infers some others, for example:

If the humidity is low, water.

Defuzzifier

In this stage, the results of the fuzzy inference, which are numerical values, are transferred to the variables that take linguistic values.

Fuzzy systems handle degrees of uncertainty, which according to Heisenberg (1927, mentioned by Carvalho, Bassanezi, & Lodwick, 2017), is described by linguists to describe certain phenomena within the sensible world in degrees that represent qualities or partial truths, in those that use sophistical language in various concepts and that serve to classify concepts such as tall, heavy, smoker or sick.

In the case of geographic location maps, they normally do not show risk areas and that is why a degree of uncertainty is needed to know if a place is dangerous and what are the variables that are involved in that perception.

Uncertainty is a perception of human reason and care must be taken so that the application of techniques that manage it produces positive solutions results to problems. The definition of uncertainty is based on the logic that:

$\chi A\colon U \to \{0, 1\}$

which means that the elements of the universal set U are also elements of A. There are some cases where an element is partially in a set, which means that it belongs to two sets, in different degrees.

Uncertainty generates difficulty in decision-making or planning tasks and more so in operations where the results could affect people or entities or even cause companies to lose money or resources.

This investigation, it is proposing a technique based on fuzzy logic in which uncertainty in degrees of danger is managed to determine, according to the data obtained in real-time, if an area is dangerous or how dangerous it is for certain crimes.

This fuzzy technique is managed uncertainty in degrees of danger is managed to determine, according to the data obtained in real-time, if an area is dangerous or how dangerous it is for certain crimes National Geographic (2022).

Maps have various benefits for people, such as the exact locations of houses, streets, or neighborhoods, or revealing where people live; It also shows how to get to certain places or even the travel time and also, and people can learn about the context of a place, such as its flora, fauna, climate or the relief it has and the limits of the different countries can be located or cities; In addition, the maps can also provide the population or economic statistics of a place, as well as its urbanizations or its natural resources.

Knowing the places and their contexts are very important in the daily life of people since they can plan if they go on vacation to the beach or not because there is a hurricane; They can know where there are volcanoes and their risk limits, or they can even show the path of a car to find out where they are in time and space.

Concatenated with the above, multiple benefits can be listed, some of which are mentioned below:

- Avoid getting lost or helping to get out of a maze that someone has gotten into when driving a vehicle.
- Locate a place with the help of GPS.

or even:

- There are applications like Google maps in which you can see a flat map, but also a three-dimensional map and you can even perceive how the streets are displayed.

Cartographers use many tools and knowledge to make maps that are increasingly useful to society; helped with various tools such as photographs that show the streets or the reliefs of the maps, too.

GPS to help locate objects, geometry to measure distances from one place to another and even to help plan a trip or to know how long it has taken a person to move from one place to another; In addition, they show the level of traffic of the places and give options with travel alternatives; But the crime rate is not shown on them, so questions like the following areas:

What if someone wants to live somewhere and wants to know the crime rate in that area?

What happens if a woman wants to visit somewhere? She would like to go by car and eat somewhere, but if she sees on the map that crime is high in terms of vehicle theft, business robbery, and femicide, she will think about taking greater precautions or choosing another place to visit, if possible.

Perhaps crime in certain places has always been dangerous, or perhaps only fame remained, or worse, some places that were not dangerous have become very frequent places of opportunity for criminals.

For all of the above, it is necessary to apply the technological tools that exist now and also propose new ones with different technologies.

The crime that exists in the world is great and varied, it can be visualized in the criminal data of the areas and, the visualization of them occurs for determined periods; to give an example, in the United States, according to the U.S. Crime Index State Rank.

The site U.S. Crime Index State Rank classifies the US states by their crime rate, where it can be seen that the states with the highest crime rate are found in Table 1.

While those who offend the least are shown in Table 2.

Table 1. States with the highest crime rate

1.	3,747	District of Columbia / 633,736
2.	2,253	South Carolina / 4,727,273
3.	2,241	New Mexico / 2,080,085

Table 2. Those who offend the least

1.	950	Vermont / 626,358
2.	956	New Hampshire / 1,321,069
3.	985	North Dakota / 704,925

Problem

The maps can show help to all of the aforementioned, however, they do not give options to know the degree of crimes that occur in the areas, and although there are criminal incidents by region, they do not show the degree of danger in the areas.

General Objective

Develop a fuzzy system for the analysis of the crime rate that occurred in some places in the Mexican Republic and whose results serve to model a map with degrees of criminal belonging which serves to provide information to people for decision-making to visit or combat crime and, in addition being a reference for the optimization of crime prevention processes of police corporations.

Solutions

It is necessary to indicate dangerous areas on maps in a diffuse way so that it is easier and faster to detect; however, the world is changing so the crime incidence zones can be high one day or very high another, or more or less dangerous one day and less in five. For this reason, a technology is needed that diffusely shows these areas.

This work proposes to show the maps with a criminal incidence so that depending on the case, it will be of importance and usefulness for those who consult it.

We will begin by saying that the fuzzy technology paradigm is based on establishing that things are measured by degrees of belonging, known as membership, which in turn generates degrees of uncertainty, that is, in this case, an area could be classified with a high rate in the case of the crime of robbery, but it could have a lower crime rate in homicides and, in turn, it could be average in other crimes.

The idea of this work is to make a map as a proposal for crime indicators, which will update its areas of criminal conflict in real-time, in such a way that it can be used for citizen and police consultation.

The fuzzy technology proposed for this research is based on taking the crime rate as input variables and depending on its fuzzy inference it returns results that are used to make decisions.

The fuzzy method that this research takes into account is Mamdani Fuzzy, which bases its equation on the rules:

If a is A1 and b is B1 then c is C1
If a is A2 and b is B2 then c is C2

This means that the sets A1, A2, B1, B2, C1, and C2 have two inputs that logically generate an output evaluated by Boolean operators (Mamdani & Assilin 1975, mentioned by Gokceoglu & Sezer, 2012).

In this system and as a proposal, the variables that are the most common in crimes of impact in the regions will be taken into account, although there are more according to the penal code of each country. The variables are introduced in a fuzzy system, following the following algorithm:

1. Introduction of variables
2. Application of rules
3. Defuzzifier
4. Implementation of the cartographic map
5. Results

Methodology

To solve the proposal the following phases were carried out:

RESEARCH

To solve the problem, the first step was to investigate the crime rates of various regions. This served to choose the variables with the greatest impact, which served as a reference for the study and became input variables to the system.

The State of Mexico, Mexico, is one of the States with the highest crime incidence; its criminal traffic light; Derived from this, the crime index chosen to be a benchmark for the proposed system was that of that State; specifically of Chalco.

The proposal is to make the crime index by zones in the municipalities, this as an example. The municipality of Chalco was chosen since it is a multivariate municipality in its crime rate; In addition, according to Cisneros (2022), the inhabitants of this municipality have conditions of vulnerability that have fostered a culture of transgression, detachment, and fear, which have been generated by the criminal violence that exists in the community.

The municipality of Chalco, State of Mexico, is a place that is located in the East of the State and Mexico City, more or less in the center of the Country. Chalco is a municipality that has rural and urban localities, the same as those displayed on the map in Figure 5.

Figure 5. Original map of Chalco
Source: Own

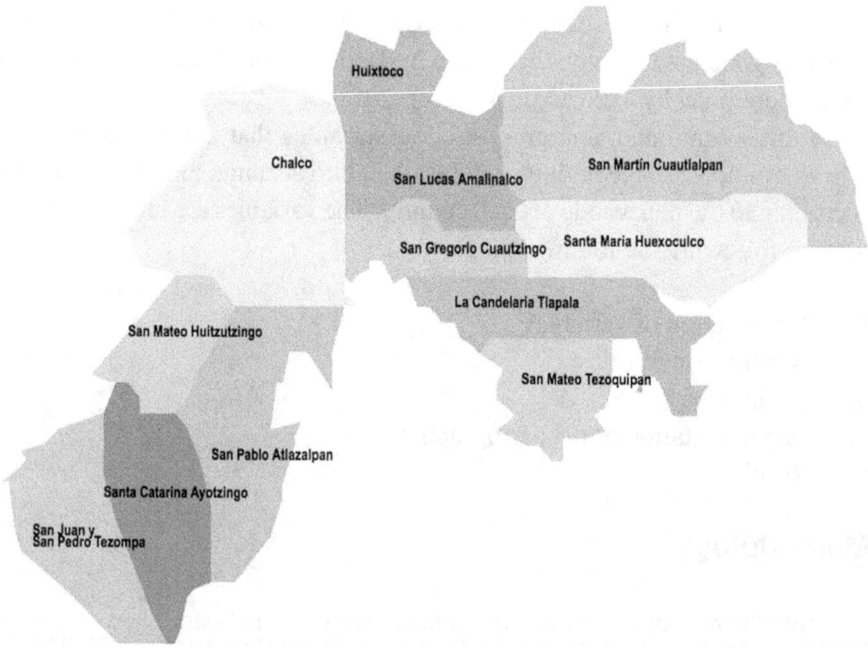

To apply the fuzzy logic in the criminal incidence, the procedure was carried out based on the crimes committed in the Municipality of Chalco, which are classified as having the greatest impact on society.

Table 3 shows the crime rate categorized into high, medium, and low; the same serves as a base to diffusely represent the degree of danger of Chalco in terms of homicide, kidnapping, extortion, drug dealing, vehicle theft, home robbery, business robbery, injuries, rape, domestic violence, and femicide.

Table 3. Criminal incidence of 2 crimes

	Universe of Discourse			
	High	**Half**	**Low**	**Criminal Incidence January 2023**
Homicide	12	6	5	5
Kidnapping	1	0	0	1
Extortion	26	8	6	10
Drug dealing	5	3	2	3
Vehicle theft	111	74	55	85
Home burglary	41	23	17	21
Business robbery	73	53	40	75
Injuries	124	110	83	91
Rape	18	9	7	4
Domestic Violence	157	83	62	83
Femicide	1	0	0	0

Source: Criminal Traffic Light, State of Mexico http://edomex.semaforo.com.mx/

The fuzzy process was carried out as follows:

Fuzzification is the process of assigning fuzzy values to sharp ones; there are different types of fuzzification; intuition is one of them and it refers to the use of human understanding to assign these values (Hoda y Raich, 2017).

The fuzzification was coded in the Python programming language, which served to determine the degree of belonging of the crimes committed in the town of Chalco, State of Mexico. The crimes were classified in ranges of danger, data that served to establish the universe of data for each crime; four of the fuzzy results are visualized in the graphs of image 6

Figure 6. Diffuse results by crime
Source: Own

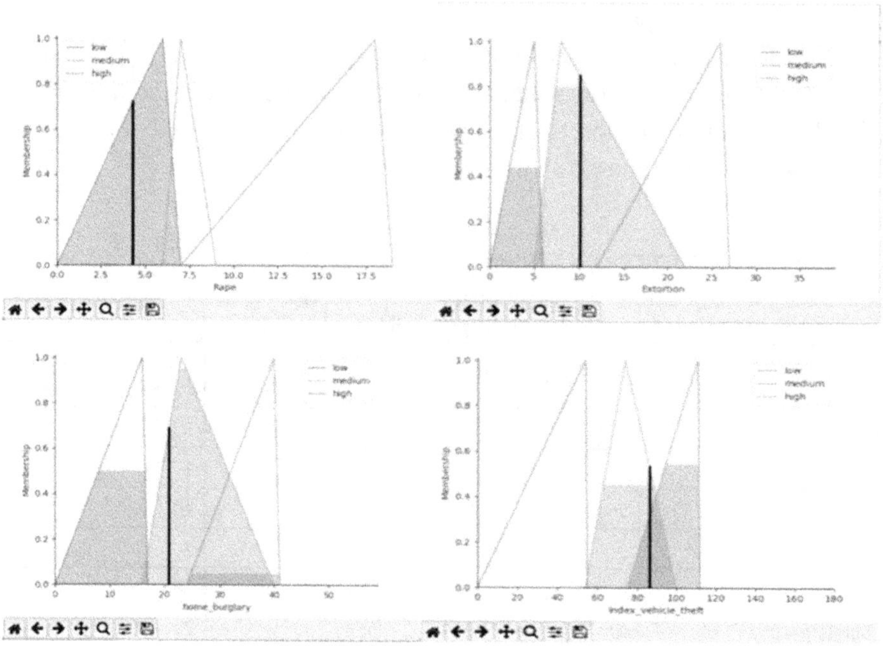

Rulebase

The conditional clauses were the criminal rates of the crimes, for example in the case of rape, the input values to the process were the low, medium, and high rates, with which the figures were labeled, obtaining the following rules:

rule1 = ctrl.Rule(A['poor'] | B['poor'], Rape['low'])

rule2 = ctrl.Rule(B['average'], Rape['medium'])

rule3 = ctrl.Rule(B['good'] | A['good'], tip['high'])

Fuzzy inference: The fuzzy inference, was made for four crimes that were selected to exemplify the process, and the result, was stated as follows:

For rape:

- If the number of violations was from 0 to 8, it was considered low
- If the number of violations was from 9 to 17, it was considered an average
- If the number of violations was 18 and over, it was considered high

Defuzzifier. In this stage, the results of the fuzzy inference, which are numerical values, are transferred to the variables that take linguistic values.

The proposal is to automate the procedure for all the towns of Chalco, but the Entity does not publish them; for this reason, it was carried out only in the town of Chalco, which is the one that is accessible. The interpretation of the results of the graphs is described below:

Rape

If the lowest crime rate is 7, the membership reaches 1, decreasing in degrees to the right or the left, tending to be less.

If the average crime index is 9, the membership reaches 1, decreasing in degrees to the right or the left, tending to be less.

If the high crime index is 18, the membership reaches 1, decreasing in degrees to the right or the left, tending to be less.

Extortion

If the lowest crime index is 6, membership reaches 1, decreasing in degrees to the right or the left, tending to be less.

If the average crime index is 8, the membership reaches 1, decreasing in degrees to the right or the left, tending to be less.

If the high crime index is at 26, the membership reaches 1, decreasing in degrees to the right or the left, tending to be less.

Home Burglary Room

If the lowest crime index is 17, membership reaches 1, decreasing in degrees to the right or the left, tending to be less.

If the average crime index is 23, the membership reaches 1, decreasing in degrees to the right or the left, tending to be less.

If the high crime index is 41, the membership reaches 1, decreasing in degrees to the right or the left, tending to be less.

Vehicle Theft

If the lowest crime index is 55, membership reaches 1, decreasing in degrees to the right or the left, tending to be less.

If the average crime index is 74, the membership reaches 1, decreasing in degrees to the right or the left, tending to be less.

If the high crime index is 11, the membership reaches 1, decreasing in degrees to the right or the left, tending to be less.

In the graphs of Figure 6, the fuzzification interprets the rates of each crime and classifies them into the categories of low, medium, and high; However, in fuzzy terms, it is observed that the universe of data has variations between crimes, since they are committed in greater and lesser intensity and, in these terms, it is possible to visualize the membership of the actual crime rate committed for each crime; In the graphs, it is visualized as a black line and is interpreted as follows:

- Violation, its membership is 0,7 categorized in low terms
- Extortion, its membership is 0,9 categorized at medium level
- Robbery to House-room, its membership is 0,7 at the medium level
- Vehicle theft with a membership of 0.58, categorized more at a high level than a medium level.

The previous idea shows that the values of the memberships are placed at the extremes of the range of the criminal index; this gives a fuzzy interpretation of the values since the crime index is located in some category to a greater or lesser degree.

If the previous premise is translated into linguistic terms, it is observed that the most widespread case is vehicle theft, in which the crime rate is high at 0.58 and medium at 0,42.

CARTOGRAPHIC SYSTEM

To represent the vagueness of the places, a map was made that shows in degrees of membership belonging to the low, medium, and high categories.

JavaScript, Html5, PHP, and MySQL; are programs used to code a system that draws a map; the task of each type of software in the fuzzy system is detailed below.

1. The coordinates of a map representing the Municipality of Chalco were obtained.
2. The map was drawn.
3. HTML software was used in conjunction with JavaScript to draw the map.
4. d) A MySQL database was created to store the crimes and their amounts; in order to automate the results shown on the map.
5. The system was coded in the PHP programming language, to connect the database with the map and thus be able to dynamically update and display the data in real-time.

After a connection was made between HTML5 and PHP to Mysql, with which a database was connected to the map, so that the colors dynamically change, depending on the crime level of the place, this derived from the graphs that the fuzzy system emits, of which variables were determined:

Danger zone
R1: If the crime is high, then the color green
R2: If the crime is medium, then color orange
R3: If crime is low, then the color blue
R4: If the offense is high and medium, then a combination of green and orange colors
R5: If the crime is medium and low, the colors are orange and blue

The rules are applied to all crimes and the results of their application are shown on the map the degree of belonging to the crime index for each crime and place. In image 7, you can see the combination of colors that are the result of the application of fuzzy rules and whose interpretation means the following:

Figure 7. Crimes committed in Chalco are seen in a diffuse way
Source: Own

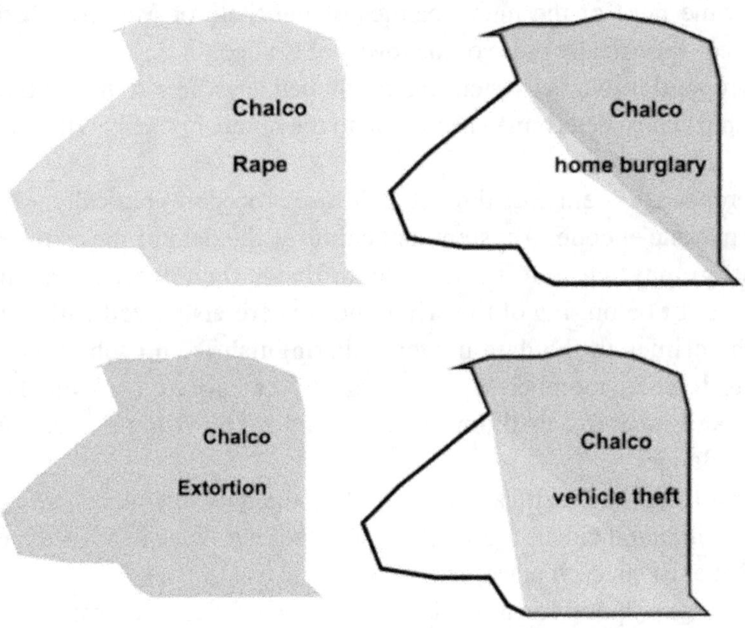

Chalco is a municipality with a low rape crime rate, which is why part of the map is painted blue.

Extortion and house robbery show a degree of belonging more to the middle state, therefore, it is displayed in more orange color.

The perception of vehicle theft shows vagueness between medium and high, therefore paint with a combination of the two colors; however, as the membership is more towards the high membership grade, the green color looks stronger than the orange.

Fuzzification converts the integer value that represents the criminal index, into a fuzzy one by assigning the degree of membership, that is, and according to what Timothy (2017) said, fuzzification is the process of converting clear quantities to fuzzy, this means to say that the crime index is fuzzy variables with the ambiguity of belonging to one place or another, which generates uncertainty and imprecision.

DISCUSSION

The fuzzy logic proposal shows that problems can be solved based on membership, which guides decision-making. In this work, the crime index located in low, medium, or high according to its membership function was shown.

The crime rates of the municipalities of the State of Mexico, Mexico were investigated; specifically those of the towns of Chalco.

The proposal is that the system can be applied anywhere so that in this way, its membership can be determined according to the number of crimes that are carried out at any given time.

The proposed system goes through 2 diffuse procedures basically:

The first one encodes a system that captures the data of the crime rates and returns its graphs that show the rates according to their degree of membership. The degrees of belonging of the crime indexes are visualized with colors; this makes the crimes in the data universe distinguish from each other, and that according to their membership, they can be categorized within the ranges, low, low and medium, medium, medium and high, or high, according to its degree of danger

Derived from the fuzzification suffered by the data, the places are mapped to show the criminal incidence in degrees of belonging, which shows the degree of danger that exists in each place.

This type of proposal can be made for each place if there is access to the criminal data; however, it was only done as an example for four crimes in the town of Chalco, since there are no other towns in the Municipality.

After investigating, applying the diffuseness, and simulating the diffuse maps, it can be determined that Fuzzy maps are suitable for solving case studies, and 3 examples are given to better understand how they could be implemented:

Case 1: If the person who consults the fuzzy maps is a woman who wishes to visit Chalco to meet someone for lunch when she consults the map, she will observe that rape is low, although vehicle theft is high, consequently, you will have to be careful in that aspect.

Case 2: If a person plans to go live in Chalco, he will surely focus on house robbery, data that is a medium trend at this time; likewise, if the is a woman, she will confidently view rape safety because it is perceived to be low.

Case 3: If a person wants to start a business, they will surely focus on business robbery or extortion, which, as shown in image 7, belongs to the medium status, which means that it is not very safe, but not the opposite either.

For all the above and concluding conjunction with research done on crime prevention or attack, the fuzzy theory could add to efforts to lower crime rates, proposing optimization of prevention processes to organizations that fight crime.

CONCLUSION

Combating and preventing crime must receive more attention from police organizations, but to a greater extent from individuals; that is, people must be careful in their walk through the world.

It is essential to devise new strategies of action for this prevention of crimes; Although there are indeed cities in the world where there are no high crime rates, at least in Latin America, insecurity is a daily issue.

There are many strategies to combat crime, but most of them are attacks, that is when the crimes have already happened; what is needed is for the population to know how to take care of themselves and, where appropriate, what to do to prevent being subject to criminal acts.

This research is a proposal for crime prevention since it proposes a system based on fuzzy logic so that governments can apply it in their localities and thereby make their regions safer places.

If the proposed systems could be made global and located like a Google Maps tool or something similar, people could prevent crime and thus reduce crime rates.

The elaboration and visualization of maps with fuzzy logic techniques can be handled by more specific places.

The fuzzy system could be more robust, if the number of crimes committed in specific places, were processed daily.

The number of crimes that are committed is captured daily by government institutions. Both the Municipal and State Police or the Justice Prosecutor's Office do it, but they only report long periods.

The system should process the data issued by the reports, but if it were done daily, the crime rate on the maps would be displayed at all times.

The fuzzy theory handles the vagueness of things; in the proposed examples, it cannot be affirmed or denied that there are secure or completely insecure sites; however, if you can point out the ones that are most insecure in certain crimes.

In the case of vehicle theft in Chalco, the map was colored green; In the case of the crimes of extortion and business robbery, the incidence was medium, therefore the map was colored more orange. For the crime of rape, the color was blue, because the incidence was low.

This research proposes a way for the population to protect itself as much as possible, about something that is already known, but not with certainty.

The uncertainty of things can be applied in the best way, to obtain results that encourage people to take care of themselves.

This proposal would warn people about the dangers they would be exposed to if they went somewhere, although it would give them confidence about others in that same place.

People would use the maps at their convenience, based on their reasons for wanting to attend a specific place.

Finally, the fuzzy theory cannot only be applied to electronic systems but can also be applied to health, manufacturing, or social issues and this is a clear example of its application.

REFERENCES

Bondy, M., Roth, S., & Sager, L. (2020). Crime Is in the Air: The Contemporaneous Relationship between Air Pollution and Crime. *Journal of the Association of Environmental and Resource Economists*, *7*(3), 555–585. doi:10.1086/707127

Cardona, C., Restrepo, D., & Ovalle, D. (2007). Sistema Experto Difuso para Determinar Perfiles Criminológicos basado en el Test de Lüscher y Variables Socio-Criminológicas. *Revista Avances en Sistemas e Informática, 4*(2), 97-105. Recuperado de: https://www.redalyc.org/articulo.oa?id=133116858012

Carvalho, de B. L., Bassanezi, R. C., & Lodwick, W. A. (2017). *A first course in fuzzy logic, fuzzy dynamical systems, and Biomathematics Theory and Applications*. Springer Berlin Heidelberg.

Cisneros, J. L. (2020). Violencia y criminalidad: El caso del Municipio de Chalco, Estado de México. *SAPIENTIAE: Revista de Ciencias Sociais, Humanas e Engenharias, 5*(2), 272-285. Recuperado de: https://www.redalyc.org/articulo.oa?id=572762223002

Demir, C., & Ozcan, S. E. (2022). Do air quality and green space reduce propensity to crime?: an empirical investigation of OECD countries. Management of Environmental Quality. doi:10.1108/MEQ-08-2022-0235

Flores Payán, L., & Refugio Vallejo, J. (2015). Evaluación de políticas y programas sociales mediante lógica difusa. *Nóesis. Revista de Ciencias Sociales y Humanidades, 24*(1), 83–114. doi:10.20983/noesis.2015.1.3

Franco, M. S., Valdez P. M. (2022). *Principios básicos de cartografía y cartografía automatizada.* UNAM.

Freilich, J. D., & Newman, G. R. (2017). Situational crime prevention. Oxford research encyclopedia of criminology and criminal justice. doi:10.1093/acrefore/9780190264079.013.3

Geografía (INEGI). (2022, January 1). *Incidencia delictiva.* Encuestas En Establecimientos. Especiales. Encuesta Nacional de Victimización de Empresas. ENVE, Encuestas En Hogares. Encuesta Nacional de Victimización Y Percepción Sobre Seguridad Pública. ENVIPE. https://www.inegi.org.mx/temas/incidencia/

Gokceoglu, C., & Sezer, E. (2012). Soft computing modeling in landslide susceptibility assessment. In *Terrigenous mass movements* (pp. 51–90). Springer. doi:10.1007/978-3-642-25495-6_3

Grant, H. (2015). The "Social" of Crime Prevention: Meaning and Implications of Social Crime Prevention for Police in the Developing World. In *Social Crime Prevention in the Developing World* (pp. 1–7). Springer. doi:10.1007/978-3-319-13027-9_1

Haczek, Á. R. (2022, September 1). *Estos son los estados de EE.UU. con más y menos delitos violentos.* CNN. https://cnnespanol.cnn.com/2022/09/01/delitos-violentos-estados-unidos-orix/#:~:text=En%20Estados%20Unidos%2C%20la%20media

Halder, A., & Kumar, A. (2019). Active learning using rough fuzzy classifier for cancer prediction from microarray gene expression data. *Journal of Biomedical Informatics*, *92*, 1–13. doi:10.1016/j.jbi.2019.103136 PMID:30802546

Hattingh, M. J. (2015). The use of Facebook by a Community Policing Forum to combat crime. *Proceedings of the 2015 Annual Research Conference on South African Institute of Computer Scientists and Information Technologists.* 10.1145/2815782.2815811

Hernández B. H. (2021). *Homicidios en América Latina y el Caribe: magnitud y factores asociados*. Academic Press.

Hoda, D. S., & Raich, V. (2017). *Fuzzy Logic Models and Fuzzy Control*. Alpha Science International LTD.

INEI. (2016). *Instituto Nacional de Estadistica e Informatica.* https://m.inei.gob.pe/prensa/noticias/en-el-ano-2016-la-tasa-de-homicidios-fue-de-77-muertes-por-cada-100-mil-habitantes-9856/

Khalid, S., Khan, S. A., & Ifzal, S. Q. (2021). A Fuzzy Logic-Based Framework for Mapping Crime Data on Established Sociological Hypothesis for Societal Disorder Identification and Prevention. *IEEE Access : Practical Innovations, Open Solutions*, *9*, 80197–80207. doi:10.1109/ACCESS.2021.3083542

Lutz, Roth, Sefi, Bondy, & Malvina. (2019). Crime is in the air: the contemporaneous relationship between air pollution and crime. Harvard Dataverse. doi:10.7910/DVN/RMMITA

National Geographic. (2022). *Map | National Geographic Society.* Education. nationalgeographic.org. https://education.nationalgeographic.org/resource/map

ONU-Habitat. (2022). *Reducción del crimen a través de la planificación y gestión urbana.* https://onuhabitat.org.mx/index.php/reduccion-del-crimen-a-traves-de-la-planificacion-y-gestion-urbana

Palmer, D., Warren, I., & Miller, P. (2018). Crime prevention, dataveillance, and the regulation of information communication technologies. In *Encyclopedia of Social Network Analysis and Mining* (pp. 463–474). Springer New York. doi:10.1007/978-1-4939-7131-2_210

Ramos-Soto, A., & Pereira-Fariña, M. (2017). On modeling vagueness and uncertainty in data-to-text systems through fuzzy sets. arXiv preprint arXiv:1710.10093.

S.A.S., E. L. R. (n.d.). *Los países con los índices más altos de criminalidad organizada en todo el mundo.* Diario La República. https://www.larepublica.co/globoeconomia/los-paises-con-los-indices-mas-altos-de-criminalidad-organizada-en-todo-el-mundo-3351148

Timothy, J. R. (2017). Fuzzy Logic with Engineering Applications. John Wiley and Sons, Ltd.

Torres, G. E. P., & Granadillo, A. (2019). *Estudio Dogmático y filosófico del concepto de delito*. J.M Bosch. doi:10.2307/j.ctvq2vx91

U.S. Crime Index State Rank. (n.d.). http://www.usa.com/rank/us--crime-index--state-rank.htm

Veerasamy, B., & Sangeetha, C. M. (2022). MB-FL: Macro-Block Fuzzy Logic for Video Compression in Multimedia Applications. *International Journal of Fuzzy Logic and Intelligent Systems*, *22*(4), 366–372. doi:10.5391/IJFIS.2022.22.4.366

Wang, C., Chen, D., & Hu, Q. (2014). Fuzzy information systems and their homomorphisms. *Fuzzy Sets and Systems*, *249*, 128–138. doi:10.1016/j.fss.2014.02.009

Zadeh, L. A. (1965). Fuzzy sets. *Information and Control*, *8*(3), 338–353. doi:10.1016/S0019-9958(65)90241-X

Zadeh, L. A. (1977). Fuzzy sets and their application to pattern classification and clustering analysis. In *Classification and clustering* (pp. 251–299). Academic Press. doi:10.1016/B978-0-12-714250-0.50014-0

Chapter 3
Design and Simulation of an Automated Sanding Machine for the Roughing of Plastic Parts

Noemi González León
https://orcid.org/0000-0001-5261-4539
Higher Technological Institute of the Sierra Norte de Puebla, Mexico

Abraham Hernández Santiago
Higher Technological Institute of the Sierra Norte de Puebla, Mexico

ABSTRACT

This chapter presents a proposal for the design and simulation of an automated sanding machine for the roughing of plastic parts for the company SOLGISTIKA. S.A. DE C.V. The methodology used is fuzzy logic; the software used for the design of the machine is Solidworks for the simulation of the Proteus electronic components and for the control of the rules of the Matlab system. This research is carried out by the Division of Electromechanical Engineering of National Technology of Mexico, Higher Technological Institute of the Sierra Norte de Puebla.

INTRODUCTION

According to José Jiménez Pineda, general director in 2009, the company SOLGISTIKA S.A de C.V was founded, starting with recruitment activities and training of personnel, to perform in-house tasks in distribution centers of international

DOI: 10.4018/978-1-6684-8639-9.ch003

logistics operators in Mexico City. The problem that the company presents is that in one of its production areas the number of parts for some air conditioning components, lights and aircraft cabins has increased considerably, therefore, it is necessary to rework in the sanding process, in the plastic injection points to readjust the parts externally since they have a greater thickness, being complicated the roughing of the piece to meet the required quality standard of the finished product; In addition, the investment of time is greater because it involves the labor of the technologist.

The aim of this study is to design a plastic sanding machine for the company SOLGISTIKA. S.A DE C.V, in order to automate the sanding process, by eliminating burrs and plastic injection points through a cutter, on the pieces to be worked for a certain time. The main benefits of the design of the machine is to present the proposal for the increase of productivity and reduction of time of the personnel in charge of the process of roughing more complex parts, with greater difficulty for external and internal rework. It is then possible to design an automated industrial machine with the specific function of operating the roughing of plastic injection points, at easily adjustable speeds.

For the design of the automated sanding machine, we will use computer numerical control (CNC), which means that the machine is capable of measuring, automating and controlling its own processes. In this way, a task that would normally have to be executed by several operators, it is possible for the machine to carry it autonomously. A CNC machine works thanks to a controller installed in the machines. The controller receives the orders through the firmware which is installed from manufacturing or through the computer, then the controller transforms the instructions received into electrical impulses that will activate the mechanical parts of the CNC machine so that it can be put to work

For the design and construction of a semi-automatic machine for sanding flat metallographic samples (Santos et al., 2003). The number of samples that can be linked in a complete process are 8, they performed sanding tests on both annealed AISI 1020 steels and subsidized AISI 4340, on 6160 aluminum alloys and phosphorus bronze, performing for each material 3 tests at load and constant speeds varying the sanding time.

For its proposal of parametric design and production in Manufacturing Laboratory (Fablab), as an alternative to reduce the emission of carbon dioxide. Case study. They occupy open source software "OpenScad", to interpret 2D and 3D maps allowing to edit the variables of the tables such as the width, location and thickness of the object in this case the furniture. The methodology used is quantitative deductive (Barreno & Monar, 2023).

For the optical evaluation of surface roughness in wood parts processed by robotic sanding; They occupy a hybrid method based on obtaining characteristics using the co-occurrence matrix, in images processed with edge detection algorithms. The algorithm they use is Roberts and the contrast characteristic are the most suitable for evaluating the surface quality in sanded wood pieces (Iglesias et al., 2022). For its design and manufacture of a CNC milling machine. It occupies laser cutting technology (Contreras, 2021).

The proposal of Huamani (2019) is analyzed because it manufactures a CNC machine to make figures in low relief. Your CNC machine manufacturing process will be verified. It does not make samples. For the refurbishment of a two-stage industrial sander for the company "Maexa S.R.L."; company engaged in the import, export, exploitation, processing and marketing of all wood raw materials; reconditions and puts into operation the two-stage industrial sanding machine. The two-stage sanding machine is used specifically for the calibration or sanding of wooden surfaces, for the sanding process the machine consists of 4 systems: electrical, mechanical, pneumatic system, programmable logic control (Condori, 2018).

Daza (2016) proposes a CNC router applied to the development of PCBs. Design and manufacture linear displacement systems for each of its axes. Manufacture a table for the main structure of the CNC Router; It adapts and programs the software, develops a control circuit and a power supply for the stepper motors. The machine has the ability to mass produce with great precision in details. The occupied bit is only able to move up and down and right and left, it does not rotate, this means that it is not able to undermine. Design and build a prototype CNC milling machine. He designed each piece individually and then made a virtual assembly of all the pieces to observe possible interferences between them and be able to redesign. The electronic design is simplified by using commercial controllers for the stepper motors that are responsible for making the movement of the machine, tested the behavior of each axis separately; Once the calibration of the three axes has been carried out, the tests were carried out in different materials (Hurtado, 2015).

Caballero (2013) recategorizes and the value of the carpentry workshop of the Naval Arsenal port Belgrano. He tells us about a high accident rate within the carpentry workshop and its impact on the workshop. Molina (2012) designs, builds, and programs a prototype of a CNC machine for milling and punching PCB printed circuit boards. It automates the process of punching printed circuit boards and elaborate a basic milling system, occupies computerized numerical control and computer-aided design. Garcia et al. (2009) designed and implemented a CNC control to create 3D models and sculptures from a CAD design. They noted that for their proposal the machine is not appropriate to occupy the USB port, because it does not remain open, a factor that prevents the sporadic transmission of data. Stepper motors are ideal for mechanisms that require very precise movements.

METHODOLOGY

The methodology used is by components by the different elements and electronic devices of the plastic sanding machine. Fuzzy logic is used for the importance and precision of the machine. SolidWorks software is used for the design of the machine. Proteus for the electronic system; Java Processing Wiring for the microcontroller firmware; MatLab for diffuse control; for simulation operations of plastic sanding machine Universal Gcode Sender.

Plastic Sanding Machine Design: Figure 1 shows the design of the sanding machine base for plastic roughing.

Figure 1. Base of the prototype of the plastic sanding machine developed in SolidWorks
Source: Own

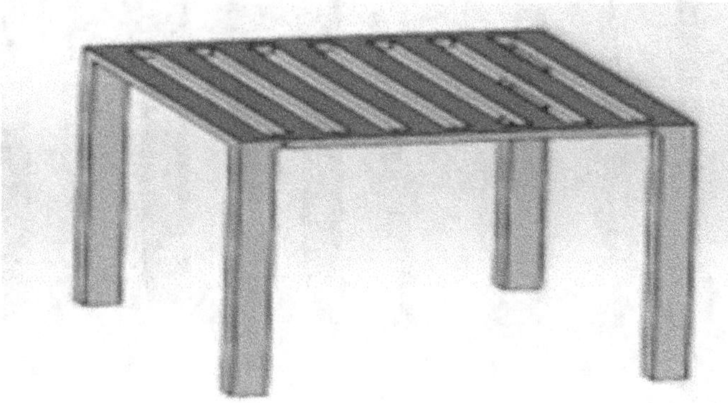

The machine must contain four supports that will form the y-axis, two square tubes of 5x5 where the rails of the mechanism move as can be seen in Figure 2.

Figure 2. Supports on the y-axis of the plastic sanding machine
Source: Own

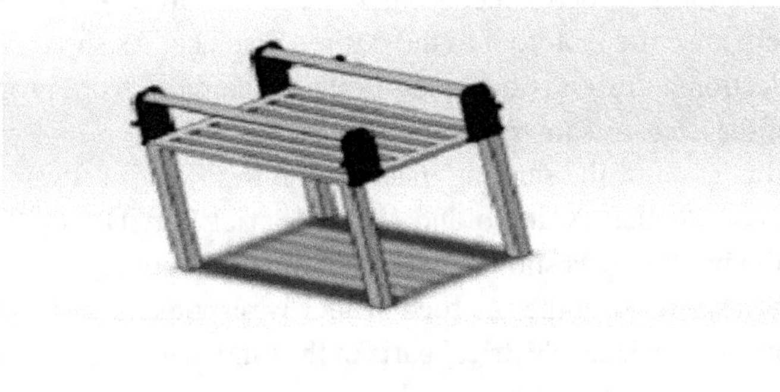

After having the supports of the y axis, a fastener is placed to hold the x-axis, to achieve displacement throughout the square tube, see Figure 3. Inside the bra was placed four tires that help move along the rail that makes up the axle. A motor should be inserted into the same fastener to help move in the set direction. Three small pulleys are placed on the outside of the mechanism to help the motor make its trajectory by means of a belt, as illustrated in Figure 3.

Figure 3. Fastener design for displacement x and outer part of the mechanism
Own elaboration.

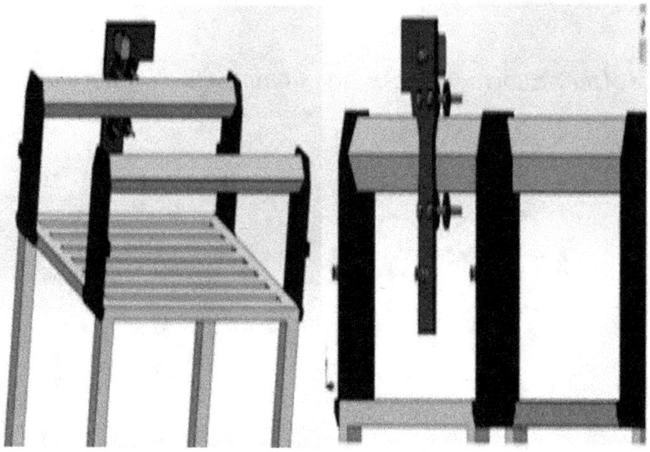

A tube is placed for the x-axis, the rims and pulleys that will make up the entire axle, two fasteners are placed so that there is no imbalance at the time of making the journey and help facilitate the mobility of the rail. For the part that holds the cutter is placed a tube three wide by twenty-five centimeters high. This will be placed between the x-axis, and the motor to control the z-axis. The z fastener is placed, in the lower part another fastener is inserted, to hold the mototool that works as a cutter, which has inputs for different types of bits, as can be seen in Figure 4.

A motor fixed to the top of the z-axis is placed, a fifteen-centimeter screw is inserted to help raise the mototool to the desired distance. As shown in Figure 5.

For the electronic design is occupied Proteus, the motors are programmed for the control of the axes, as shown in Figure 6.

For the simulation of the sanding machine the following software is occupied: Universal Gcode Sender, Arduino and Virtual Serial Port Driver. The ports are configured for simulation as shown by Port Com5 in Figure 7.

The firmware housed in the microcontroller is responsible for communication between the mechanical and electrical parts of the sander and the computer; to carry

Figure 4. Tube for the z-axis and fastener of the mototool
Source: Own

Figure 5. Screw attached to the z-axis motor
Source: Own

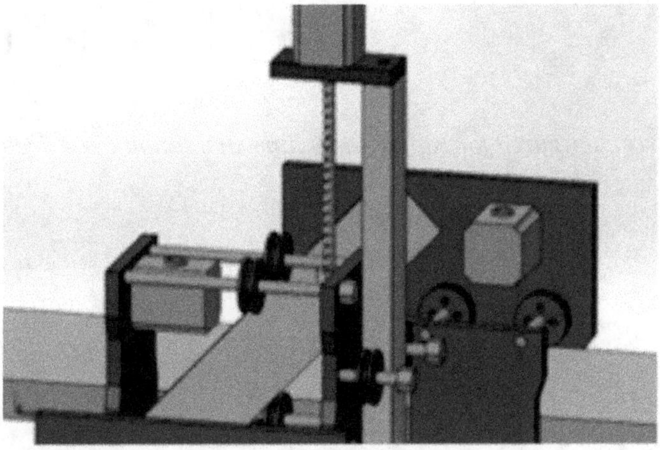

out and achieve the simulation of the system it is necessary that it is linked; the microcontroller program with Universal Gcode Sender. Therefore, it is essential to verify if you have the software loaded, in the case of not having it, it is advisable to compile the program and if it does not present errors, load the file to the microcontroller and verify the data thrown on the serial monitor.

With the Virtual Serial Port Driver, it is verified if the communication port is installed for the operation of the microcontroller, later, we will relate it to Universal Gcode Sender.

Figure 6. Connection diagram for PCB and operation of sanding machine shafts

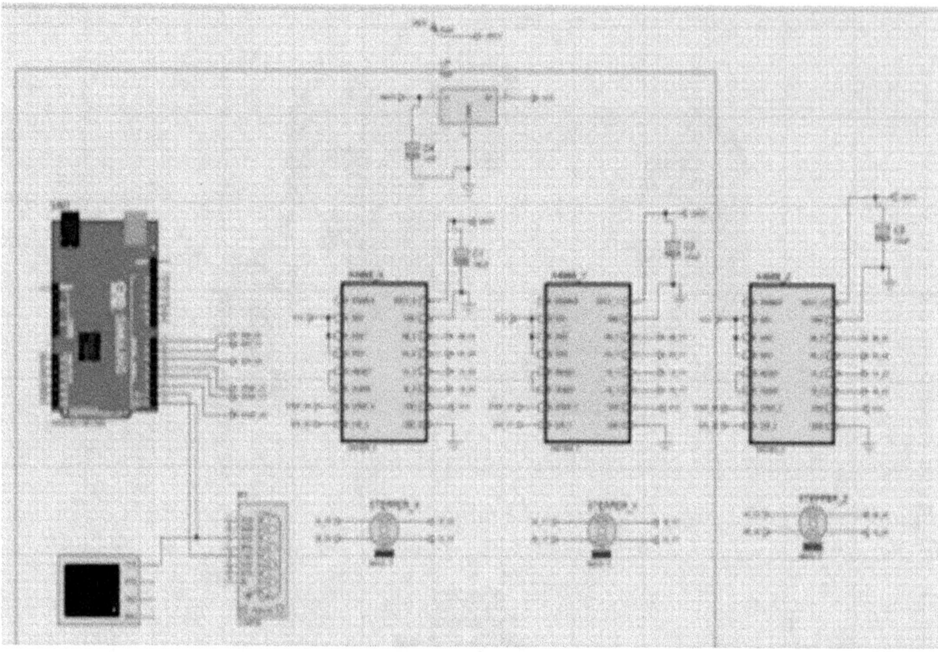

Figure 7. Ports configured for sanding machine simulation in device manager

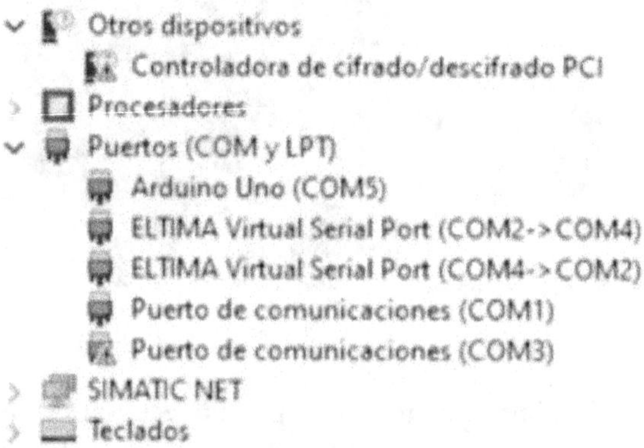

To simulate the operation of the sanding machine is occupied the *Universal Gcode Sender* program, the COM5 port and the 115200 baud are enabled, as shown in Figure 8 the bauds and measurements placed for the axis and as well as the direction to work are being recognized.

Figure 8. Simulation of the operation of the sanding machine with the Universal Gcode Sender

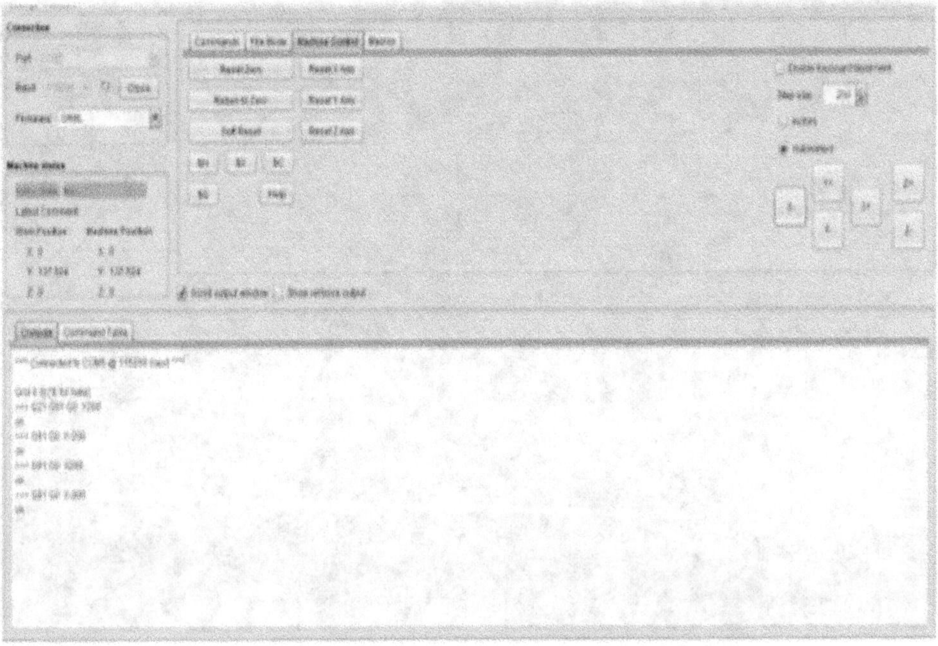

To verify that the data thrown by the simulator are correct, the MSDOS operating system is used as shown in Figure 9.

RESULTS

To obtain the results of the Fuzzy System it was necessary to occupy Matlab, the busy editor is mamdani; One of the input data of the system to control the sanding machine is the variable speed, as shown in Figure 10.

Figure 10 shows the control for the state level of the variable speed.

For the output variables of the work done by the machine we have roughing, polishing and polishing, as can be seen in Figure 11.

In the inference engine we find three fundamental rules of the control system which are:

1. If the speed is low of the machine the work of the part is only a roughing.
2. If the speed is medium the work of the piece is polished.
3. If the speed is high the work of the piece is a polish.

Figure 9. Verification of the instructions sent by the Universal Gcode Sender to the MSDOS operating system.

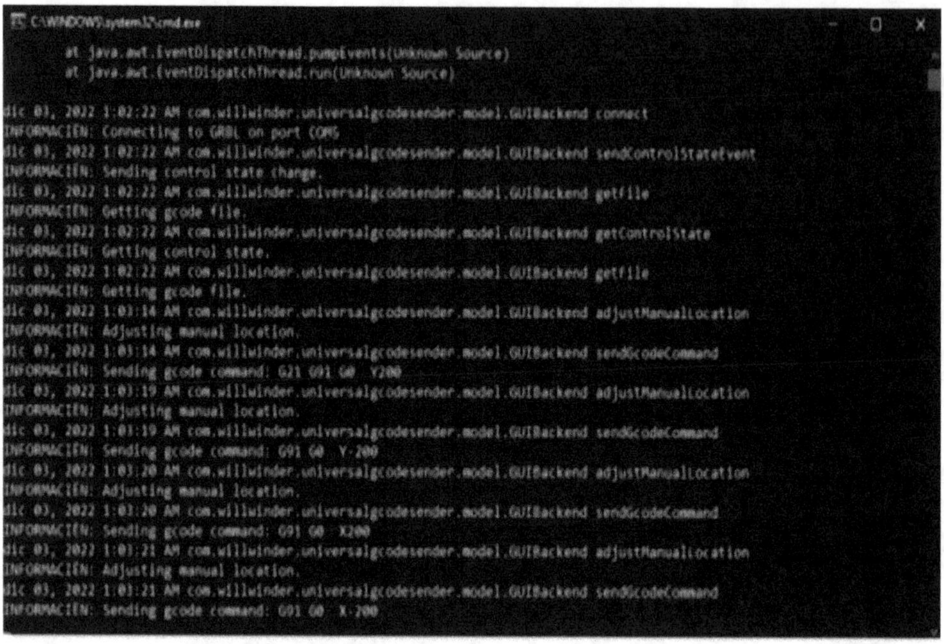

Figure 10. Control of the input variable: speed in a low, medium, and high state

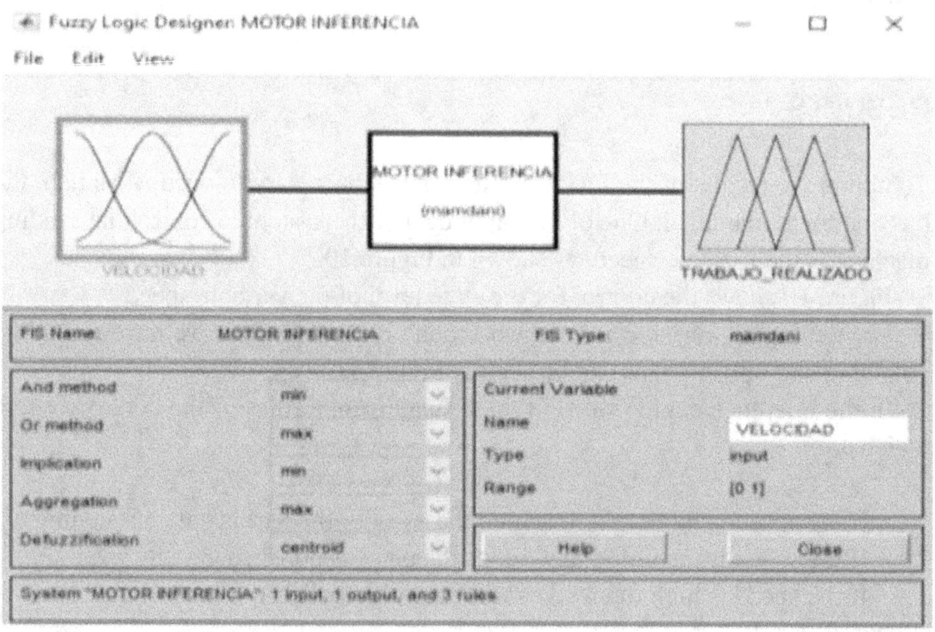

Figure 11. Work done by the plastic sanding machine.

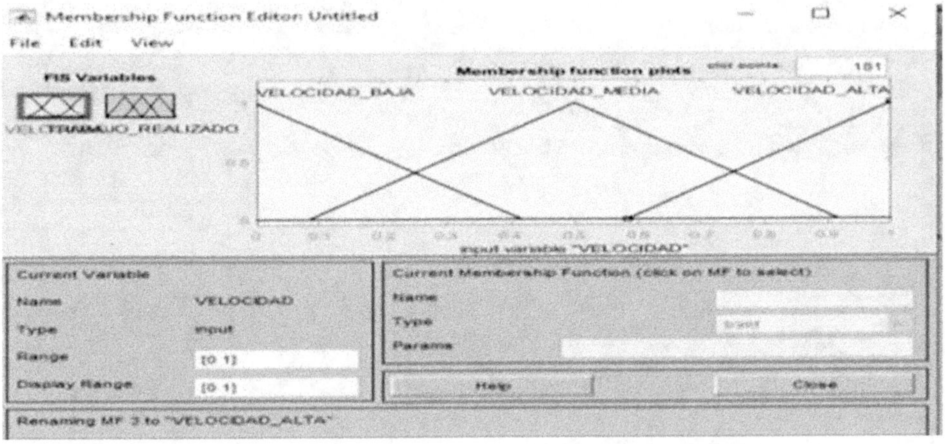

Figure 12 shows how fuzzy rules are declared within the system.

Figure 13 shows the Fuzzy Inference Diagram, the process of involvement from its beginning to the end of it. In it you can move the line indices that corresponds to the inputs and then see how the system readjusts and calculates the new output. If the speed is 0.5, the output of the work done by the sanding machine is observed.

Figure 12. Diffuse rules

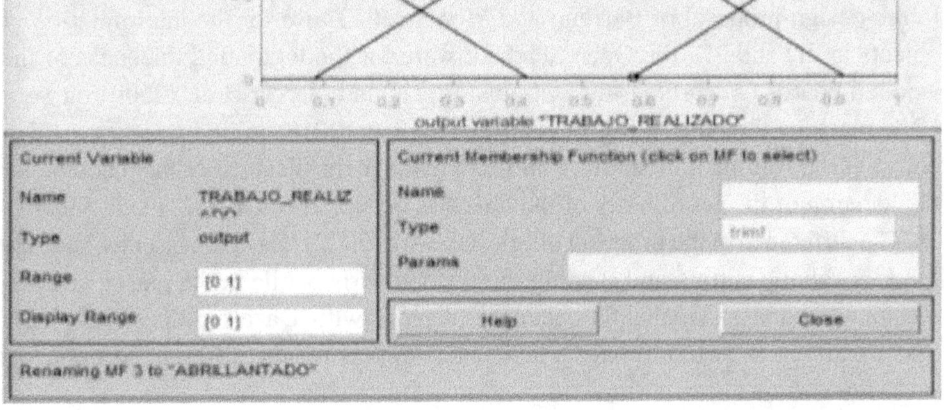

Figure 13. Fuzzy inference diagram of plastic sanding machine

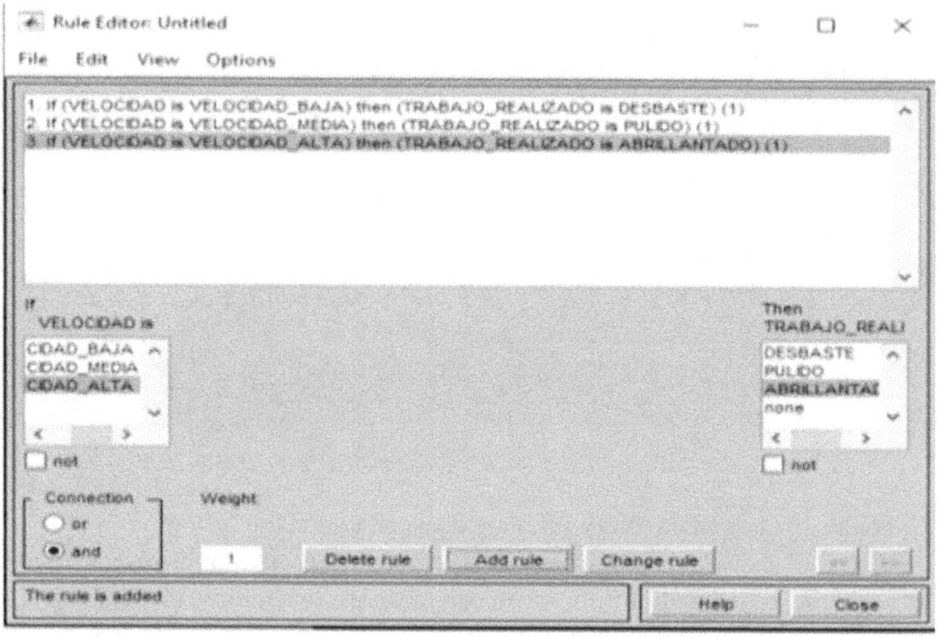

The resulting sigmoidal membership function represents the parameters located at the end of the curve; The speed and work performed on the sanding machine can be found in Figure 14.

DISCUSSION

The design of the semi-automatic machine proposed by Santos et al. (2003). It is for the sanding of steels while our proposal is for the roughing of plastic parts. In the design proposal of Barreno and Monar (2023) use for the interpretation of objects in 2D and 3D, the Open Scad software for the width and thickness of the wooden boards. According to Contreras (2021) and García et al. (2009) on your CNC machine the busy software is *CAD*. Condori (2018) in its two-stage sander, occupies the Automation Studio, Fluid Sim while in the plastic sanding machine for the location of the coordinates of the object is occupied *Universal Gcode Sender*.

The difference of the proposal of Iglesias et al. (2022) lies in the laser technology used for robotic cutting and sanding on wooden parts; while in the plastic sanding machine a band sandpaper is occupied. Agrees with Daza's proposal in 2016. Specifically in the need to have a table as a support for the different components

Figure 14. Membership function: The sigmoidal represents the work done and the speed of the sanding machine

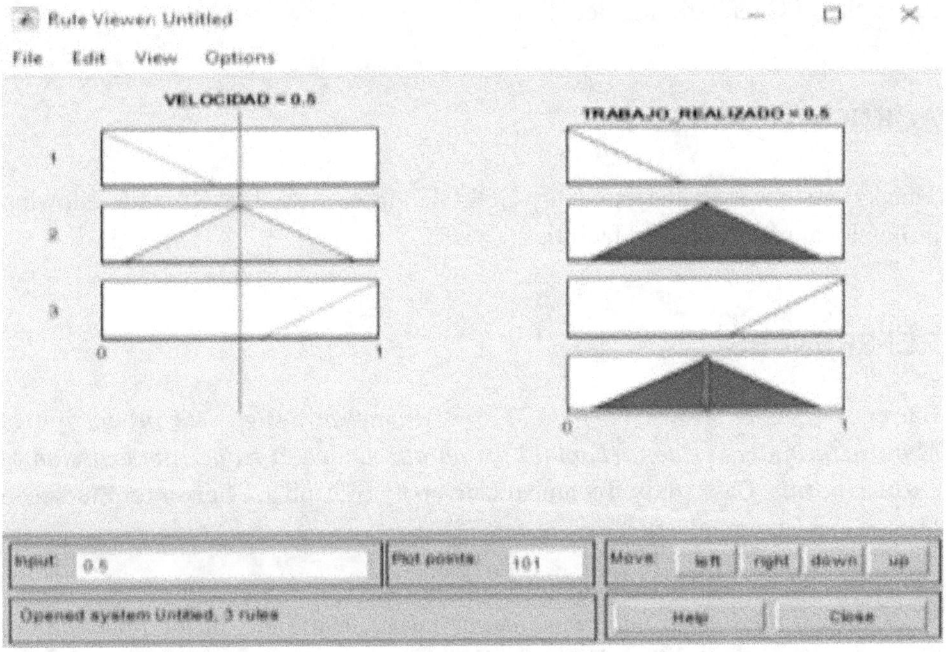

of the plastic sanding machine. Like Hurtado (2015), each of the axes of the CNC machine is tested, in this study each of the axes of the plastic sanding machine are tested at the simulation level.

CONCLUSION

A plastic sander has been designed in *Solidworks*, the connection diagram is elaborated in *Proteus*, for the Electronic System; The firmware of the microcontroller is developed in *Java Processing Wiring*, for the mechanical operation of the axes the operation of the x, y, z axes has been simulated in *Universal Gcode Sender* in combination with the firmware of the microcontroller; for the diffuse control system MatLab was taken care of; where the inference engine is established with the essential rules governing the control of the fuzzy system of the machine. The membership function is obtained with the parameters of speed and work performed for the correct operation of the plastic sander. Therefore, the objective proposed at the beginning of the research has been met, this has been achieved through the elaboration of the design of the Automated Plastic Sanding Machine and the study of the current situation in the company SOLGISTIKA. S.A DE C.V, which is located in the city of Querétaro, Mexico.

ACKNOWLEDGMENT

Thanks are given to the company SOLGISTIKA. S.A. DE C.V. for allowing professional practices in its facilities.

REFERENCES

Barreno A. A. & Monar N. M. (2023). *Parametric design and production in Manufacturing Laboratory (Fablab), as an alternative to reduce the emission of carbon dioxide*. Case study. Technical University of Ambato. Center for Studies in Design and Communication.

Caballero, H. A. (2013). *Recategorization and enhancement of the Carpentry Workshop of the Puerto Belgrano Naval Arsenal. University of the Fraternity of Groups St. Thomas Aquinas*. Faculty of Engineering.

Condori, V. C. (2018). Reconditioning of a two-stage industrial sander for the company "MAEXA S.R.L." Universidad Mayor de San Andrés. Faculty of Technology. Electromechanical Career.

Contreras, A. J. A. (2021). *Design and manufacture of a CNC milling machine. University of Jaén. Higher Polytechnic School of Linares*. Department of Mechanical Engineering.

Daza, M. U. R. & Luengas, Z. S. (2016). *CNC Router Applied to the Development of PCB's*. National Polytechnic Institute. School of Mechanical and Electrical Engineering Professional Unit Adolfo López Mateos.

Garcia, M. I. C., Lagos, L. J. G., Urrego, P. L. F., & Delgado, P. P. Y. (2009). Design and implementation of a CNC control to create models and sculptures in third dimension from a CAD design. University of San Buenaventura. Faculty of Engineering. Mechatronics Program. *Bogotá D., C*, 190.

Huamani, B. R. P. (2019). *It manufactures a CNC machine to make figures in low relief. National University of San Agustin de Arequipa. Faculty of Process Engineering. Academic Department of Metallurgy Engineering and Environmental Engineering*. School of Metallurgical Engineering.

Hurtado, J. C. (2015). *Design and construction of a CNC-milling machine prototype*. Carlos III University of Madrid.

Iglesias, Q. F., Aguilera, L. T., Padilla, B. A., Madaf, P. M., & Ten, C. E. (2022). *Optical evaluation of surface roughness in wood parts processed by robotic sanding*. Department of Mechanical Engineering, Faculty of Engineering and Sciences, University of La Frontera, Temuco, Chile and Forest Products Laboratory, Institute of Forests and Society, Faculty of Forest Sciences and Natural Resources, Universidad Austral de Chile, Valdivia, Chile.

Molina P. A. F. (2012). *Design, construction and programming of a prototype of a CNC machine for milling and punching PCB printed circuit boards*. University of Nariño. Faculty of Engineering. Electronic Engineering Program.

Roses. S. (2019). *Logistics. Solgistika starts operations in Queretaro*. somosindustria.com

Santos, L. N., Lopez, C. J. E. A., & López, C. J. C. (2003). Design and construction of a semi-automatic machine for sanding flat metallographic samples. University of El Salvador. Faculty of Engineering and Architecture School of Mechanical Engineering.

Chapter 4
Revolutionizing Smart Contracts:
Overcoming Challenges and Proposing Sustainable Solutions

Rinat Galiautdinov
https://orcid.org/0000-0001-9557-5250
Independent Researcher, Italy

ABSTRACT

The current study examines the challenges associated with smart contracts and their overall concept and offers a resolution that can effectively address all the prevailing issues. Moreover, it advocates for a shift in the current architecture employed in smart contracts and blockchain. The study identifies the problems, conducts an analysis, and presents a solution that expounds on how the proposed changes can enhance the performance of smart contracts based on a new architecture, boost their scalability, change the IoT world, and promote an eco-friendly environment by reducing pollution of natural environment and conserving energy.

INTRODUCTION

For more than a decade, blockchain has been recognized as a technology that serves as a distributed database (Bernstein & Lange, 2016) used for keeping track of financial transactions involving various types of cryptocurrencies, such as BTC and ETH (Daniel R. L. Brown, 2016). All payments or financial transactions are conducted through a peer-to-peer network, which decentralizes the network

DOI: 10.4018/978-1-6684-8639-9.ch004

participants and grants them equal privileges from the network's perspective. This network architecture differs from the regular client-server architecture wherein all clients rely on the server. If the server is down, no client can connect to it and conduct any type of transaction (Heuvel, 2014).

The introduction of the blockchain technology also introduced a new term, "smart contracts," which were initially proposed in the 1990s by Nick Szabo (1997). The concept is simple: smart contracts are computer programs that verify different conditions and determine which function(s) to perform based on those conditions.

A typical schema of blockchain is shown in Figure 1.

Figure 1. A typical schema of blockchain

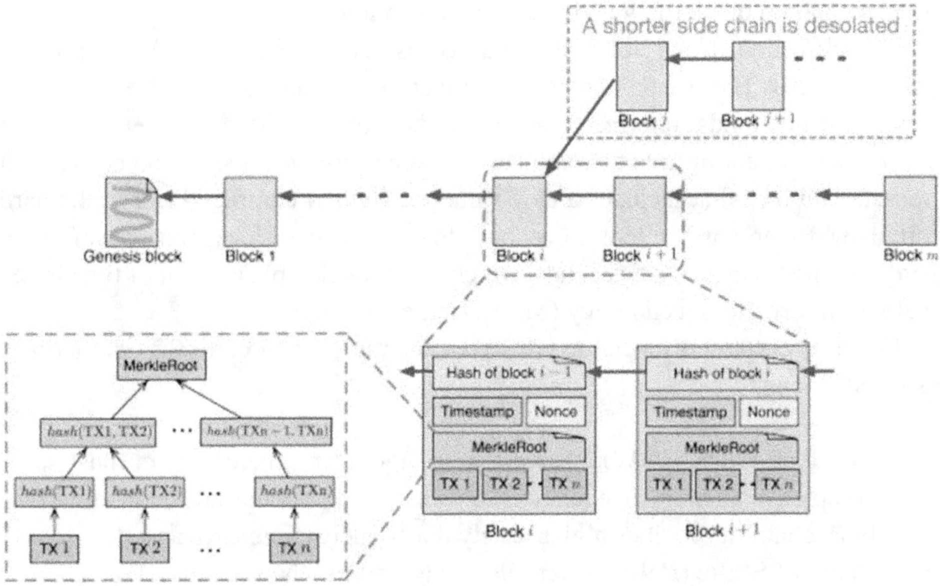

The first widely recognized cryptocurrency was Bitcoin, despite the fact that the first one emerged in the early 1990s (Li et al., 2017). The Bitcoin blockchain solely contains financial transactions and no other data. ETH, the second well-known cryptocurrency in this field, introduced the concept of "smart contracts" to the blockchain world. A smart contract can be viewed as a small program that runs on a blockchain when it receives a request (Nguyen, 2019). Smart contracts enable users to add additional logic not only to financial transactions but also to create simple games. From a technical perspective, a smart contract is regarded as a user that can receive and send transactions. This "user" maintains a record of various states, known as variables, which can be modified and/or read when stored in a blockchain. Although reading blockchain data is free for users, it remains quite sluggish.

The sluggishness of reading data from a blockchain can be attributed to various factors, one of which is the sheer size of the blockchain. As more data is added to the blockchain, the processing and verification requirements also increase, which can result in longer transaction times and slower data retrieval. Additionally, the consensus mechanism employed by the blockchain can have an impact on the speed of data retrieval. For instance, some consensus mechanisms, like proof-of-work, mandate that nodes in the network perform intricate mathematical computations, which can be time-consuming and impede the overall performance of the blockchain. While the decentralized nature of the technology also contributes to the sluggishness, the scalability and speed of blockchain technology are being enhanced through the implementation of sharding and other scaling solutions.

The alteration of variables, also known as state change, is a costly process (Zhang, 2020). The reason behind this is that every state change necessitates the expenditure of funds, referred to as gas in the Ether world. Moreover, it takes a substantial amount of time to complete the transaction, process it, and construct a transaction block that includes a new value for a smart contract. Prior to the hard fork in the Ether world, it took 21 seconds to create a new block, and even after the hard fork, it still takes 12.5 seconds, which is a considerable amount of time in the realm of information technology (Mkrttchian et al., 2019).

Different crypto-currencies use different approaches to process the transactions, as listed below:

- **Proof of Work (PoW):** A consensus algorithm where miners must solve complex mathematical equations to validate transactions and add them to the blockchain. It provides high security but is energy-intensive and slow.
- **Proof of Stake (PoS):** A consensus algorithm where validators must hold a certain amount of cryptocurrency to validate transactions and add them to the blockchain. It is energy-efficient and faster than PoW but provides lower security.
- **Practical Byzantine-Fault Tolerance (PBFT):** A consensus algorithm that allows a network of nodes to reach a consensus even if some nodes are malicious or faulty. It is fast and efficient but requires a high number of nodes.
- **Low Level Virtual Machine (LLVM):** A compiler infrastructure that allows developers to write programs in high-level languages and compile them into machine code. It is widely used in the development of operating systems, programming languages, and other software.

- **Convolutional Neural Network (CNN):** A type of deep learning algorithm used in image and video recognition, natural language processing, and other applications. It recognizes patterns in data and learns from them.
- **Long Short-Term Memory (LSTM):** A type of recurrent neural network used in speech recognition, language translation, and other applications. It remembers long-term dependencies in data and avoids the problem of vanishing gradients.
- **Ether (ETH):** The cryptocurrency used in the Ethereum blockchain. It is used to pay for transaction fees, smart contract execution, and other services on the network.
- **Bitcoin (BTC):** The first and most well-known cryptocurrency. It is used as a store of value and a medium of exchange.
- **Ethereum Virtual Machine (EVM):** A virtual machine that runs smart contracts on the Ethereum blockchain. It is designed to execute any program written in any programming language.
- **Unspent Transaction Output (UTXO):** A record of the unspent output of a transaction. It prevents double-spending and ensures the integrity of the blockchain.
- **Internet of Things (IoT):** A network of interconnected devices that exchange data and communicate with each other.
- **Distributed Autonomous Corporation (DAC):** An organization that is run by smart contracts on a blockchain. It is decentralized, autonomous, and transparent.
- **Certificate Authority (CA):** An entity that issues digital certificates to verify the identity of users, devices, and organizations on the internet.
- **Delegated Proof of Stake (DPOS):** A consensus algorithm where token holders vote for delegates who validate transactions and add them to the blockchain. It is faster than PoW and PoS but provides lower security.
- **WebAssembly (Wasm):** A binary format that allows developers to write programs in any programming language and run them in web browsers. It is designed to be fast, efficient, and secure.
- **Border Gateway Protocol (BGP):** A protocol that exchanges routing information between different networks on the internet. It ensures efficient and secure data transmission.

An illustration of a smart contract between a buyer and a supplier can be used as an example. As depicted in Figure 2, the supplier initiates the process by sending a product catalog to the buyer through the blockchain network. This catalog contains product descriptions, shipping details, and payment terms, and is stored and distributed on the blockchain. This enables the buyer to access the product information and

Figure 2. Demonstrates a interaction between a buyer and a supplier using a smart contract

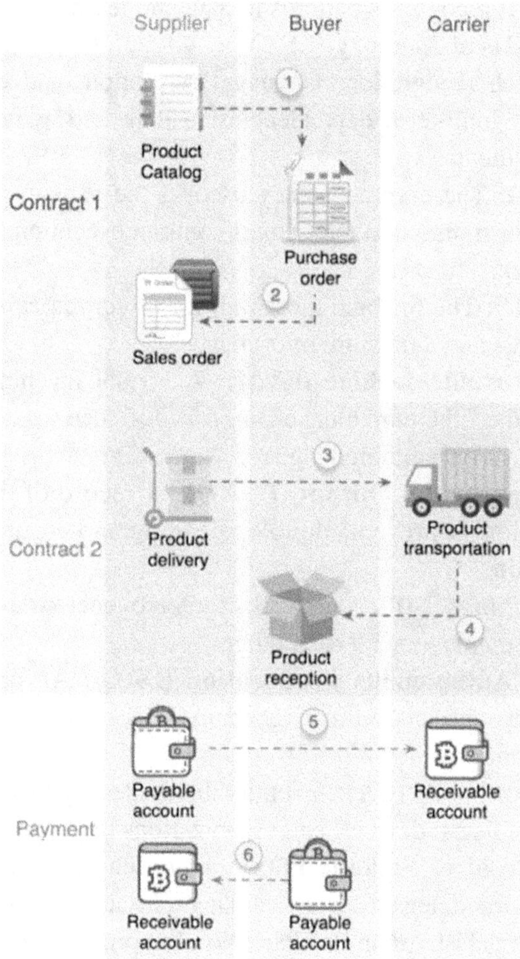

authenticate the supplier's reputation simultaneously. The buyer then places the order, specifying the quantity and payment date, using the blockchain. This entire process constitutes a purchase contract (Contract 1), highlighted in a blue box in Figure 2. It is noteworthy that the entire process is conducted solely between the buyer and supplier, without any third-party intervention.

The following contains the reasons for the popularity and usage of smart contracts:

- **Risks Optimization.** Due to conditional "immutability" of blockchains, smart contracts "cannot be" arbitrarily altered once they are issued (these issues will be discussed later on in the article). Moreover, all the transactions that are stored and duplicated throughout the whole distributed blockchain system are traceable and auditable. And finally, the result of all this:any malicious behaviors like financial frauds in theory can be greatly mitigated (however this is not always true).

- **Cutting Down Administration and Service Costs.** Block-chains assure the trust of the whole system by distributed consensus mechanisms without going through a central broker or a mediator. Smart contracts stored in blockchains can be automatically triggered in a decentralized way. Consequently, the administration and services costs due to the intervention from the third party can be significantly saved.
- **Improving the Efficiency of Business Processes.** The elimination of the dependence on the intermediary can significantly improve the efficiency of business process. Take the aforementioned supply-chain procedure as an example. The financial settlement will be automatically completed in a peer-to-peer manner once the predefined condition is met (e.g., the buyer confirms the reception of the products). As a result, the turnaround time can be significantly reduced.
- **Transparency:** It happens by monitoring the transactions between the parties undertaking the contract in a blockchain. This ensures there is transparency between the transactions.
- **Manages Transactions.** As mentioned earlier, parties or companies cannot change the contract's agreement terms once it has gone live. This is useful in handling different transactions in the blockchain. Fast processing of transactions occurs since smart contracts utilize simple logic to perform that. Also, users are assured that their claims will be processed faster, leading to quick payouts.
- **Minimal Paperwork.** Smart contracts ensure there will be less paperwork involved as it does not need the lawyers to set up an agreement. Also, less time is involved as no pre-contract talks are involved in the agreements.

Upon completion of Contract 1, the supplier will utilize the blockchain to locate a carrier for the shipping phase. Like Contract 1, the carrier will also publish shipping details, including transportation fees, source, destination, capacity, and shipping time, along with conditions and terms in the blockchain. If the supplier accepts the carrier's contract, the products will be delivered to the carrier who will then dispatch them to the buyer, resulting in the creation of Contract 2 (enclosed in the pink box) as illustrated in Figure 2. The entire process of Contract 2 is also conducted without the involvement of a third party.

In addition to the automatic execution of Contract 1 and Contract 2, payment procedures, including payment from the supplier to the carrier and payment from

Figure 3. The number of smart contracts created in Ethereum blockchain

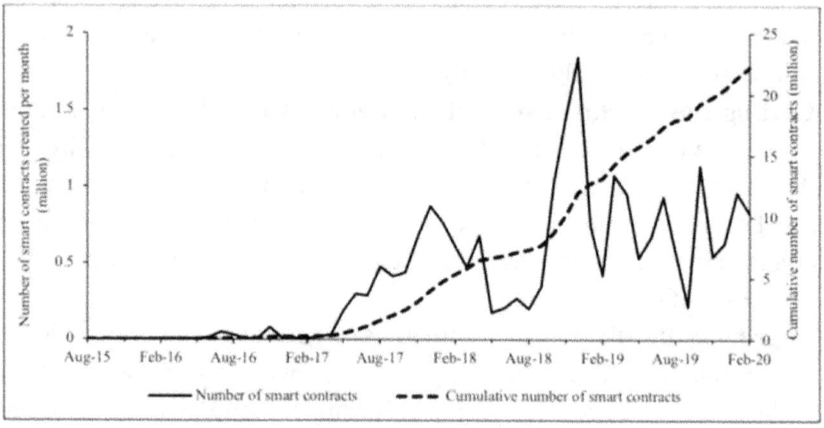

the buyer to the supplier, are also completed automatically. For instance, once the buyer confirms the receipt of the products, payment between the buyer and supplier is automatically triggered as per the predefined condition. The financial settlement from the buyer to the supplier is conducted using crypto currencies such as Bitcoin or Ether.1. Unlike traditional transactions, the entire process is carried out in a peer-to-peer manner without the intervention of third parties like banks, resulting in significant savings in turnaround time and transactional costs.

Here is a comparison of various Smart Contract Platforms:

1. Ethereum
   - First smart contract platform
   - Widely adopted and established community
   - Supports Solidity programming language
   - High transaction fees and scalability issues
2. EOS
   - High transaction speed and scalability
   - Uses Delegated Proof of Stake (DPoS) consensus algorithm
   - Limited decentralization due to block producers
   - Vulnerable to censorship and governance issues
3. TRON
   - High transaction speed and scalability
   - Uses Delegated Proof of Stake (DPoS) consensus algorithm
   - Focuses on entertainment and gaming industry
   - Limited decentralization due to block producers

4. Cardano
    - Uses Proof of Stake (PoS) consensus algorithm
    - Focuses on academic research and scientific approach
    - High scalability and security
    - Still in early development stages
5. NEO
    - Supports multiple programming languages including C#, Java, and Python
    - Focuses on digitizing real-world assets
    - Uses Delegated Byzantine Fault Tolerance (dBFT) consensus algorithm
    - Limited scalability and adoption
6. Hyperledger Fabric
    - Enterprise-focused platform
    - Permissioned blockchain with high privacy and security
    - Supports smart contracts in multiple programming languages
    - Limited adoption outside of enterprise use cases
7. RSK
    - Smart contract platform built on top of Bitcoin blockchain
    - Uses merge-mining to secure network
    - Supports Solidity programming language
    - Limited adoption and scalability
8. Qtum
    - Hybrid blockchain platform combining Bitcoin and Ethereum technologies
    - Uses Proof of Stake (PoS) consensus algorithm
    - Supports smart contracts in multiple programming languages
    - Limited adoption and scalability
9. Chainlink
    - Decentralized oracle network for smart contracts
    - Provides external data and information to smart contracts
    - Highly secure and reliable
    - Limited functionality beyond oracle services.

Smart contracts are also subjected to attacks, so in 2016 when the Decentralized Autonomous Organization (DAO) smart contract was manipulated to steal around 2 Million ETH (50 Million USD on the time) because of its re-entrancy vulnerability (Tasca, 2019). In addition to the vulnerability problem, smart contracts face several challenges including privacy, legal, and performance issues.

Figure 4. One of the examples of re-entrancy attacks on a smart contract

```
// SPDX-License-Identifier: MIT

pragma solidity ^0.8.0;

contract Auction {
    mapping(address => uint) public bidders;
    bool locked;

    function bid() external payable {
        bidders[msg.sender] += msg.value;
    }


    function refund() external {
        uint refundAmount = bidders[msg.sender];

        if (refundAmount > 0) {
            (bool success,) = msg.sender.call{value: refundAmount}("");
            require(success, "failed!");
            bidders[msg.sender] = 0;
        }
    }

    function currentBalance() external view returns(uint) {
        return address(this).balance;
    }
}
```

From the first glance at the smart contract the refund method has to get the value of the funds of a user and send him back his funds: *(bool success,) = msg.sender.call{value: refundAmount}("");*

After which the line *bidders[msg.sender] = 0;* marks that his current balance is 0.

And it seems to be ok and it was a regular application (not a smart contract of Etherium) it would work fine. But not in the world of Etherium.

What happens here is: an intruder first calls for a *proxyBid()* function to send some funds to the Action contract. Then the Action contract will receive this money and will make a record that a new user made a bid equal to some amount of the money. But then intruder calls for a *attack()* function which automatically calls a *refund()* function of the Action smart contract. The Action smart contract checks whether the user has the funds (yes) and tries to send them: *(bool success,) = msg.sender.call{value: refundAmount}("");* However in the world of smart contract of Etherium instead of making an update on the user's record making it 0, it will run the *receive()* function of an intruder which contains another request for the Action smart contract: *action.refund()* and as a result, the intruder will get all the money possessed by Auction smart contract.

For developers who entered the world of smart contracts from other areas of software development, it may not be immediately apparent.

Figure 5. Another smart contract which makes a re-entrancy attack on the auction contract

```
contract AuctionAttack {
    uint constant BID_AMOUNT = 1 ether;
    AuctionHacked auction;


    constructor(address _auction) {
        auction = AuctionHacked(_auction);
    }

    function proxyBid() external payable {
        require(msg.value == BID_AMOUNT, "incorrect");
        auction.bid{value: msg.value}();
    }

    function attack() external {
        auction.refund();
    }

    receive() external payable {
        if(auction.currentBalance() >= BID_AMOUNT) {
            auction.refund();
        }
    }

    function currentBalance() external view returns(uint) {
        return address(this).balance;
    }
}
```

Figure 6. This attack can be avoided if we make some changes in the action smart contract

```
    modifier noReentrancy() {
        require(!locked, "no reentrancy!");
        locked = true;
        _;
        locked = false;
    }

    function refund() external noReentrancy {
        uint refundAmount = bidders[msg.sender];

        if (refundAmount > 0) {
            bidders[msg.sender] = 0;

            (bool success,) = msg.sender.call{value: refundAmount}("");

            require(success, "failed!");
        }
    }
```

Functional issues with current smart contract platforms are numerous, and several representative challenges are presented in this paper. Firstly, re-entrancy, which allows the interrupted function to be safely recalled again, can be exploited by malicious users to steal digital currency, as indicated in Li et al. (2017). Secondly, block randomness is required by certain smart contract applications, such as lotteries and betting pools. While pseudo-random numbers can be generated in a block timestamp or nonce, malicious miners may fabricate blocks to deviate from the outcome of the pseudo-random generator, thereby controlling the probability distribution of the outcomes, as shown in Bonneau et al. (2015). Lastly, recent work has revealed that smart contracts can be overcharged due to under-optimization, with features such as dead code and expensive operations in loops consisting of repeated computations, as demonstrated in Chen et al. (2017).

Below are most often advances for functional issues:

- **Re-Entrancy:** Recently, several proposals attempt to solve some of the above challenges. Obsidian was proposed to address re-entrancy attacks and money leakage problems. In particular, Obsidian exploits *named* states to enable consistency checking on state transitions and verification so that re-entrancy vulnerability can be mended. Moreover, a data flow analytical method was proposed to prevent the illegal digital currency stealing from the leakage.
- **Block Randness:** Blockchain is regarded as a promising technology to generate public and unpredictable random values. However, the random output might not be so random as people expect. Miners could control the block generation and release the block until they find it profitable. To address this issue, some approaches proposed to use the delay-function to generate randomness. It means that the random value will be only be known to others after a short time period since its generation. In this way, the blockchain moves on and the miners could not withhold their blocks to profit. But delay functions are not suitable for smart contracts as most of they require instance verification. To this end, the approach proposed the Sloth function to allow faster verification. Based on above propositions a multi-round protocol to verify delay functions using a refereed delegation model. It reduces the cost of verifying the output from $30 to $0.4 (Tessaro, 2016).
- **Overcharging:** Besides from caring the efficiency of their programs, developers of smart contract also need to pay attention to their execution costs. It's noticed that over 90% of real smart contracts suffer from gas-costly patterns in Ethereum (Zhang et al., 2019). A researcher proposed GasReducer, a tool used to detect gas-costly patterns. GasReducer can replace under-optimized byte code with efficient byte code.

Smart contracts encounter more than just various types of attacks as their problems are not restricted. Their potential is significantly reduced due to numerous limitations. Furthermore, smart contracts are now perceived as a component of a payment system (such as in the ETH blockchain), which is illogical since the processing of financial transactions should not be associated with the blockchain that stores data. The current structure adversely affects pollution, transaction costs, and energy expenditure, all of which could be readily addressed.

LITERATURE REVIEW

For several years, there has been a debate in the business world regarding the potential benefits and limitations of smart contracts. These contracts are self-executing and contain the terms of agreement between the buyer and seller in lines of code. They have the potential to increase efficiency, reduce costs, and prevent fraud. However, complex contracts that require human interpretation and discretion may not be suitable for this technology, and there is a risk of hacking or programming errors.

According to the Harvard Business Review, smart contracts are most suitable for simple, standardized agreements, such as those used in supply chain management. The article highlights the need for significant investment in technology infrastructure and expertise, which could limit their widespread adoption. Additionally, legal and regulatory frameworks are crucial to support the use of smart contracts, especially in areas like contract enforcement and dispute resolution.

The TechRadar piece delves further into the advantages and drawbacks of smart contracts. While they can decrease transaction costs and enhance efficiency by removing intermediaries, the article cautions that smart contracts may be susceptible to hacking or programming mistakes, potentially causing significant losses for contract parties. The article concludes that organizations must assess the potential benefits and drawbacks of smart contracts before adopting them.

Similarly, the CIO Dive article highlights the limitations of smart contracts. It argues that they may not be appropriate for all contract types, especially those requiring human interpretation and discretion. The article also warns that smart contracts may be vulnerable to hacking or programming errors, leading to significant losses for contract parties. The article recommends that organizations carefully weigh the potential benefits and drawbacks of smart contracts before integrating them.

Overall, the articles suggest that while smart contracts have the potential to offer significant benefits, they may not be suitable for all types of contracts. Organizations should carefully consider the potential benefits and drawbacks of smart contracts before implementing them, and should ensure that legal and regulatory frameworks are in place to support their use. Additionally, organizations should invest in technology infrastructure and expertise to ensure that smart contracts are implemented correctly and securely.

PROBLEM

The existing approach to the architecture of the smart contracts provides lots of disadvantages, such as:

1. **Immutability:** Once a smart contract is deployed on the blockchain, it cannot be changed or modified. This means that any errors or bugs in the code cannot be fixed, and any unintended consequences of the contract cannot be rectified.
2. **Complexity:** Smart contracts can be complex and difficult to understand, especially for non-technical users. This can lead to errors in the code or misunderstandings of the contract terms.
3. **Lack of Legal Recognition:** Smart contracts are not yet legally recognized in many jurisdictions, which can create uncertainty around their enforceability and validity.
4. **Dependence on Blockchain Technology:** Smart contracts are dependent on the underlying blockchain technology, which can be vulnerable to attacks or technical issues. This can lead to the loss of funds or other negative consequences.
5. **Cost:** Smart contracts can be expensive to deploy and execute, especially on public blockchains where transaction fees can be high.
6. **Limited Functionality:** Smart contracts are currently limited in their functionality and cannot perform complex tasks or interact with external systems without the use of additional tools or protocols.
7. **Lack of Privacy:** Smart contracts are typically public and transparent, which can be a disadvantage in situations where privacy is important.
8. **Difficulty in Debugging:** Debugging smart contracts can be difficult and time-consuming, especially for complex contracts. This can lead to errors in the code or unintended consequences of the contract terms.
9. **Lack of Flexibility:** Smart contracts are designed to be self-executing and autonomous, which can limit their flexibility in certain situations where human intervention may be necessary.
10. **Scalability:** Smart contracts can be limited in their scalability, especially on public blockchains where transaction throughput can be low. This can limit their usefulness in high-traffic applications.

To begin with, let us revisit the scenario involving Luna, ETH, and BTC blockchain. Additionally, developers have the ability to include multiple versions of the final smart contracts, which may not be evident to an inexperienced user. Due to the fact that smart contracts are often unalterable (as we have seen in instances where ether and other currency blockchains were reverted because of lack of profitability for certain groups), developers utilize a programming technique that enables them to alternate between various versions of smart contracts.

As an example, such the implementation could be done in the following way:

Are there other advantages of placing the smart contracts to blockchain? Highly likely the answer is No.

Figure 7.

```
//SPDX-License-Identifier:MIT

pragma solidity ^0.8.13;

contract Proxy {

    address public implementation;
    uint public x;

    function setImplementation(address _imp) external {
        implementation = _imp;
    }

    receive() external payable {
    }


    function _delegate(address _imp) internal virtual {
        assembly {
            calldatacopy(0, 0, calldatasize())

            let result := delegatecall(gas(), _imp, 0, calldatasize(), 0, 0)

            returndatacopy(0, 0, returndatasize())

            switch result
            case 0 {
                revert(0, returndatasize())
            }
            default {
                return (0, returndatasize())
            }
        }
    }

    fallback() external payable {
        _delegate(implementation);
    }
}


contract V1 {
    address public implementation;
    uint public x;

    function inc() external {
            x+=1;
    }

    function enc() external pure returns (bytes memory) {
        return abi.encodeWithSelector(this.inc.selector);
    }
```

At the same time the list of disadvantages is quite big and additionally it includes the following:

- Slow speed

A smart contract is running on a miner's machine and consequently is additionally limited by the speed of its computer.

- Unable to horizontally scale

A smart contract can't be horizontally scaled, which means a miner's machine can't create multiple instances of a smart contract to speed up its work and/or to accept multiple incoming requests. At the same time a smart contract can't be stateless (if it's doing something meaningful) and it also creates the additional issues because currently there is no way to split data from logic.

- Can't have multiple run time environment for the running Smart contracts

A smart contract does not support a situation where multiple smart contracts run in their own (isolated) runtime environments

A smart contract does not support creation of a separate processes and/or even threads which eventually have a negative effect on smart contract's performance.

- SOLID

SOLID is a very well-known paradigm in software development where S stands for Single Responsibility, meaning that each service should be responsible only for its own business entity. However this approach should be applied to the crypto-currency sphere and it demonstrates that smart contracts should never be a part of blockchain.

- Can't use external services, databases, etc.

Smart contracts can't use any external services, databases, etc. and all this dramatically limits the abilities of smart contracts.

Thus all these issues clearly demonstrate the poor side of smart contracts and the absence of any progress in this direction.

Among of the technical issues, smart-contracts based on the current architecture have negative effect on the pollution of natural environment.

One of the biggest disadvantages of smart contracts is the amount of electricity they require to operate. Smart contracts are powered by blockchain technology, which requires a large amount of computational power to process transactions. This computational power is provided by a network of computers that are constantly running complex algorithms to verify and validate transactions. This process is known as mining, and it requires a significant amount of energy.

The energy consumption of mining is a major concern for the environment. The energy required to power the network of computers that run smart contracts is significant, and it is estimated that the energy consumption of the blockchain industry is equivalent to that of a small country. This has a significant impact on the environment, as the energy required to power these computers is primarily generated by fossil fuels.

The use of fossil fuels to generate energy has a number of negative effects on the environment. Fossil fuels are a finite resource, and their extraction and use contribute to a number of environmental problems, including air and water pollution, deforestation, and climate change. The use of fossil fuels also contributes to the release of greenhouse gases, which are a major contributor to global warming.

The environmental impact of smart contracts is not limited to their energy consumption. The production and disposal of the hardware required to run smart contracts also has a significant impact on the environment. The production of hardware requires the use of a number of natural resources, including metals and minerals. The extraction of these resources can have a significant impact on the environment, including deforestation, soil erosion, and water pollution.

The disposal of hardware also has a significant impact on the environment. Electronic waste, or e-waste, is a major environmental problem. E-waste is generated when electronic devices are discarded, and it can have a number of negative effects on the environment. E-waste contains a number of toxic substances, including lead, mercury, and cadmium, which can leach into the soil and water and cause serious health problems.

Speaking of the problems related to the current architecture of smart-contracts, it's necessary to consider IoT too. The Internet of Things (IoT) is rapidly growing, and the need for secure and efficient communication between devices is becoming increasingly important. Smart contracts have been proposed as a solution to automate these communications and ensure secure transactions. However, running smart contracts on blockchain has limitations such as scalability, high transaction fees, and slow transaction times. Therefore, running smart contracts on cloud platforms can be a viable solution to overcome these limitations.

SOLUTION

After considering all relevant issues, it is apparent that smart contracts should be divided into two parts: data and services. This principle is commonly used in software development, such as in the Visitor pattern which operates on Data objects. The same principle applies in the Smart Contract domain, where splitting the contract is necessary, but not sufficient (Galiautdinov & Mkrttchian, 2019a).

To address the aforementioned challenges related to smart contracts, the next step is to remove smart contract logic from the blockchain. Storing smart contract logic in the blockchain is a mistake and a dead end for technology. The best solution is to store the logic in cloud platforms such as AWS or Azure (Galiautdinov & Mkrttchian, 2019b).

In conclusion, the new terms for smart contracts and their features should be introduced. Smart contracts can now be classified into two types: Data Contracts and Service Contracts. A Service Contract must be stateless and cannot contain any state that needs to be stored within the contract. This allows for horizontal scaling of the Service Contract to process multiple requests simultaneously.

Doing so we'll allow to resolve the following issues:

- Slow speed

Smart contract will become fast since it's running on a cloud platform.

- Unable to horizontally scale

A smart contract could easily become horizontally scaled, it will also make developers to make smart contract services stateless.

- Can't have multiple run time environment for the running Smart contracts

Each smart contract will have its own run time environment, there will be the possibilities to use threads or light threads (similar to Go Routins in Golang), separate processes, etc.

- SOLID

The extrapolation of SOLID principals will make blockchain and financial system more logical and stable.

- Can't use external services, databases, etc.

The Service contracts could easily use any type of external databases relational and/or non-relational. Which will allow to use big data and apply it to blockchain.

- Expensive to run a smart-contract

Running a smart contract will become cheap, it will cost the same as the cost of an ordinary request to a web service. The result of this will be a positive effect reflecting on natural environment and saving the energy.

- Very limited number of the operation inside of a contract

A Smart contract running on a cloud platform as a service will be able to extend its functionality and remove the limits on the number and the types of allowed operations.

- Limited types

Having a smart contract running on a cloud platform will also allow to introduce the new types which now can't be used.

The sole reason for opposing this approach is based on the belief that it could render smart contracts changeable, thus compromising the assurance that the logic remains unaltered (Galiautdinov, 2020). While this may be true, the author previously demonstrated a pattern utilized by smart contract developers, which allows for the application of varying logic based on the smart contract version, rendering any potential changes inconsequential. Consequently, with this approach, the blockchain will solely comprise Data Contracts and the version number of the Service Contract utilized for data processing. The Service Contract will be located on a Cloud Platform, have its own version, and operate on the principles of Proof-of-Stake. Proof-of-Stake is a consensus mechanism for processing transactions and generating new blocks within a blockchain. This mechanism validates entries into a distributed database and ensures the security of the blockchain, which is referred to as a database in the context of cryptocurrency.

Speaking of POS benefits:

- With proof-of-stake (POS), cryptocurrency owners validate block transactions based on the number of staked coins.
- Proof-of-stake (POS) was created as an alternative to Proof-of-work (POW), the original consensus mechanism used to validate a blockchain and add new blocks.
- While PoW mechanisms require miners to solve cryptographic puzzles, PoS mechanisms require validators to hold and stake tokens for the privilege of earning transaction fees.
- Proof-of-stake (POS) is seen as less risky regarding the potential for an attack on the network, as it structures compensation in a way that makes an attack less advantageous.
- The next block writer on the blockchain is selected at random, with higher odds being assigned to nodes with larger stake positions.

Furthermore, as previously mentioned, the implementation of POS will have a positive impact on energy conservation and pollution reduction, which is particularly crucial in today's climate. It is important to compare POS with POW, as both consensus mechanisms play a vital role in synchronizing blockchain data, validating information, and processing transactions. While each method has its advantages and disadvantages, they have distinct approaches. Validators are responsible for creating blocks under PoS, whereas miners are responsible for solving cryptographic numbers to verify transactions under PoW. To become a block creator under PoS, one only needs to own enough coins or tokens, while miners under PoW must invest in equipment and incur high energy costs. The expensive equipment and energy costs under PoW limit access to mining and enhance blockchain security. In contrast, PoS blockchains require less processing power to validate block information and transactions, reduce network congestion, and eliminate the incentive-based reward system of PoW blockchains.

Table 1. Proof

Proof of Stake	Proof of Work
Block creators are called validators	Block creators are called miners
Participants must own coins or tokens to become a validator	Participants must buy equipment and energy to become a miner
Energy efficient	Not energy efficient
Security through community control	Robust security due to expensive upfront requirement
Validators receive transactions fees as rewards	Miners receive block rewards

However even this is not the end of the story as crypto-currencies contain the number of the other weak sides which sometimes are not obvious.

The fundamental principle of crypto-currency relies heavily on hash functions, which, as per numerous accounts, have a very low probability of experiencing collisions (or more accurately, having very few collisions during a certain period). However, there is no publicly available evidence that proves this concept. Additionally, a majority of crypto-currencies, such as BTC, have an automatic self-destruct mechanism programmed into them, which could potentially lead to the downfall of the crypto world.

DISCUSSION

The proposed alternative approach is to run smart-contracts on cloud platforms on servers rather than on the blockchain. This approach has several benefits, the major of them are:

1. **Scalability:** Running smart-contracts on cloud platforms on servers enables the execution of complex contracts that require a large number of computations. This is because cloud platforms have a much higher processing power than the nodes in the blockchain network.

2. **Cost:** Running smart-contracts on cloud platforms on servers is much cheaper than running them on the blockchain. This is because cloud platforms offer much lower fees for processing transactions.
3. **Speed:** Running smart-contracts on cloud platforms on servers is much faster than running them on the blockchain. This is because cloud platforms can process transactions much faster than the nodes in the blockchain network.
4. **Increased Functionality:** Running smart-contracts on cloud platforms on servers enables the use of programming languages that are not limited by the constraints of the blockchain. This enables the development of smart-contracts with increased functionality.

Interpretation of Results

The proposed alternative approach of running smart-contracts on cloud platforms on servers provides several benefits over the current approach of running smart-contracts on the blockchain. The benefits include increased scalability, decreased cost, increased speed, and increased functionality. These benefits make smart-contracts more useful for a wider range of applications. The current approach of running smart-contracts on the blockchain has several limitations that limit their usefulness. These limitations include limited scalability, high cost, slow speed, and limited functionality. These limitations make smart-contracts less useful for a wider range of applications.

Analysis of the New Approach

The new approach of running smart-contracts on cloud platforms on servers has several advantages over the current approach of running smart-contracts on the blockchain. The advantages include increased scalability, decreased cost, increased speed, and increased functionality. These advantages make smart-contracts more useful for a wider range of applications. The new approach enables the execution of complex contracts that require a large number of computations, which is not possible with the current approach. The new approach also enables the use of programming languages that are not limited by the constraints of the blockchain, which enables the development of smart-contracts with increased functionality. The new approach also enables the processing of transactions much faster than the nodes in the blockchain network, which is not possible with the current approach. The new approach is also much cheaper than the current approach, which makes smart-contracts more accessible to a wider range of users.

And making the comparison of how the new proposed architecture change effects on the smart-contracts in the frame of IoT we can see that cloud platforms offer several advantages over blockchain for running smart contracts. Cloud platforms have higher scalability, faster transaction times, and lower transaction fees compared to blockchain. Additionally, cloud platforms offer more flexibility and customization options, allowing developers to tailor smart contracts to specific IoT use cases.

Running smart contracts on cloud platforms can also improve the security of IoT devices. By using cloud-based smart contracts, IoT devices can communicate with each other securely without the need for a central authority. This can prevent tampering with data and ensure the integrity of transactions.

Moreover, cloud platforms can provide better accessibility to smart contracts for developers. Since cloud platforms have more resources and are easier to manage than blockchain, developers can create and deploy smart contracts faster and more efficiently. This can accelerate the development of IoT applications and reduce the time to market.

CONCLUSION

The author of the article carried out a study and put forward a resolution that can tackle the prevalent problems in the present smart contract notion and entirely modify the framework utilized in smart contracts and blockchain. The author identified the commonly known problems connected to smart contracts, scrutinized them, and offered a solution that elucidates how it will transform the scenario and generate a favorable outcome not just on the effectiveness and scalability of smart contracts but also on the ecosystem by diminishing pollution and conserving energy.

In conclusion, the proposed alternative approach of running smart-contracts on cloud platforms on servers provides several benefits over the current approach of running smart-contracts on the blockchain. The benefits include increased scalability, decreased cost, increased speed, and increased functionality. These benefits make smart-contracts more useful for a wider range of applications. The current approach of running smart-contracts on the blockchain has several limitations that limit their usefulness. These limitations include limited scalability, high cost, slow speed, and limited functionality. These limitations make smart-contracts less useful for a wider range of applications. The new approach of running smart-contracts on cloud platforms on servers overcomes these limitations and provides several advantages over the current approach.

In conclusion, while smart contracts offer a number of benefits, they also have a number of disadvantages that can have a significant impact on the environment. The energy consumption of smart contracts is a major concern, as it requires a significant amount of electricity that is primarily generated by fossil fuels. The production and disposal of the hardware required to run smart contracts also has a significant impact on the environment. As the use of smart contracts continues to grow, it is important to consider the environmental impact of these contracts and take steps to minimize their negative effects. This includes exploring alternative energy sources and improving the recycling and disposal of electronic waste.

Smart contracts are a promising technology for IoT, but running them on blockchain has limitations that can hinder their adoption. Running smart contracts on cloud platforms can overcome these limitations and provide several advantages such as scalability, faster transaction times, lower transaction fees, improved security, and better accessibility for developers. Therefore, the concept of smart contracts on cloud platforms should be explored further to accelerate the development of IoT applications and enable secure and efficient communication between devices.

REFERENCES

Azimi, R., & Alcaraz, C. (2020). A blockchain-based smart contract for secure data sharing in healthcare. *Journal of Medical Systems*, *44*(1), 14. doi:10.100710916-019-1521-9

Bernstein, D. J., & Lange, T. (2016). *SafeCurves: Choosing safe curves for elliptic-curve cryptography.* https://safecurves.cr.yp.to/

Bonneau, J., Clark, J., & Goldfeder, S. (2015). On bitcoin as a public randomness source. IACR Cryptol., 1015.

Brown. (2016). *SECG SEC 1: Elliptic Curve Cryptography (Version 2.0).* https://www.secg.org/sec1-v2.pdf

Chen, T., Li, X., Luo, X., & Zhang, X. (2017). Under-optimized smart contracts devour your money. *Proceedings of 24th International Conference on Software Analysis, Evolution and Reengineering*, 442–446. 10.1109/SANER.2017.7884650

Galiautdinov, R. (2020). *Brain machine interface: the accurate interpretation of neurotransmitters' signals targeting the muscles. International Journal of Applied Research in Bioinformatics*. doi:10.4018/IJARB.2020010102

Galiautdinov, R., & Mkrttchian, V. (2019a). Math model of neuron and nervous system research, based on AI constructor creating virtual neural circuits: Theoretical and Methodological Aspects. In V. Mkrttchian, E. Aleshina, & L. Gamidullaeva (Eds.), *Avatar-Based Control, Estimation, Communications, and Development of Neuron Multi-Functional Technology Platforms* (pp. 320–344). IGI Global. doi:10.4018/978-1-7998-1581-5.ch015

Galiautdinov, R., & Mkrttchian, V. (2019b). Brain machine interface – for Avatar Control & Estimation in Educational purposes Based on Neural AI plugs: Theoretical and Methodological Aspects. In Avatar-Based Control, Estimation, Communications, and Development of Neuron Multi-Functional Technology Platforms (pp. 320–344). IGI Global. doi:10.1016/j.comnet.2021.108011

Gao, Y., Chen, Y., & Chen, K. (2021). Blockchain and smart contract-enabled secure and efficient data sharing for 5G networks. Computer Networks, 196. doi:10.1016/j.comnet.2021.108011

Heuvel, S. (2014). *The demand for short-term, safe assets and financial stability: Some evidence and implications for central bank policies*. Academic Press.

Li, W., Andreina, S., Bohli, J.-M., & Karame, G. (2017). Securing Proof-of-Stake Blockchain Protocols. In J. Garcia-Alfaro, G. Navarro-Arribas, H. Hartenstein, & J. Herrera-Joancomartí (Eds.), *Data Privacy Management, Cryptocurrencies and Blockchain Technology* (pp. 297–315). Lecture Notes in Computer Science. Springer International Publishing. doi:10.1007/978-3-319-67816-0_17

Li, X., Jiang, P., Chen, T., Luo, X., & Wen, Q. (2017). A survey on the security of blockchain systems. *Future Generation Computer Systems.*

Li, X., Li, Y., Li, X., & Li, C. (2021). A novel smart contract-based secure data sharing framework for industrial internet of things. *Journal of Industrial Information Integration*, *21*, 100177. doi:10.1016/j.jii.2021.100177

Li, Y., Li, X., Li, C., & Li, X. (2021). A smart contract-based secure data sharing scheme for cloud-assisted healthcare systems. *Journal of Medical Systems*, *45*(2), 22. doi:10.100710916-018-1142-z PMID:33426569

Liu, Y., Wu, Y., & Zhang, W. (2021). A smart contract-based secure and efficient data trading scheme for vehicular ad hoc networks. *Journal of Network and Computer Applications*, *180*, 102934. doi:10.1016/j.jnca.2021.102934

Mkrttchian, V., Gamidullaeva, L., & Galiautdinov, R. (2019). *Design of Nano-scale Electrodes and Development of Avatar-Based Control System for Energy-Efficient Power Engineering: Application of an Internet of Things and People (IOTAP) Research Center. International Journal of Applied Nanotechnology Research.* doi:10.4018/IJANR.2019010104

Nguyen, C. T., Hoang, D. T., Nguyen, D. N., Niyato, D., Nguyen, H. T., & Dutkiewicz, E. (2019). Proof-of-Stake Consensus Mechanisms for Future Blockchain Networks: Fundamentals, Applications and Opportunities. *IEEE Access : Practical Innovations, Open Solutions*, *7*, 85727–85745. doi:10.1109/ACCESS.2019.2925010

Pearce, S. (2021, June). *Smart contracts: The good, the bad and the ugly.* TechRadar. https://www.techradar.com/news/smart-contracts-the-good-the-bad-and-the-ugly

Ren, Y., Chen, H., & Lu, Y. (2020). A smart contract-based approach for secure data sharing in vehicular ad hoc networks. *Journal of Parallel and Distributed Computing*, *135*, 32–40. doi:10.1016/j.jpdc.2019.08.014

Saleh, F. (2021). Blockchain without Waste: Proof-of-Stake. *The Review of Financial Studies, 34*(3), 1156–1190. doi:10.1093/rfs/hhaa075

Staff, H. B. R. (2021, July). Smart Contracts: Hype vs. Reality. *Harvard Business Review*. https://hbr.org/2021/07/smart-contracts-hype-vs-reality

Szabo, N. (1997). *The idea of smart contracts.* Nick Szabo's Papers and Concise Tutorials.

Tasca, P., & Tessone, C. J. (2019). A Taxonomy of Blockchain Technologies: Principles of Identification and Classification. *Ledger, 4*. doi:10.5195/ledger.2019.140

Tessaro, S. (2016). Sloth: A Lightweight Delay Function for ASIC-Resistant Proof-of-Work Based on Nakamoto's Consensus Algorithm. *Journal of Cryptology*, *29*(2), 613–638.

Vogel, P. S. (2021, May). *Why smart contracts may not be smart after all.* CIO Dive. https://www.ciodive.com/news/smart-contracts-limitations/599279/

Wang, Y., Wang, L., & Wang, X. (2020). A smart contract-based approach for secure data sharing in cloud storage. *Journal of Network and Computer Applications*, *168*, 102553. doi:10.1016/j.jnca.2020.102553

Xu, J., Li, X., & Li, Y. (2019). Smart contract-based secure data sharing scheme for cloud-assisted industrial Internet of Things. *Journal of Industrial Information Integration, 15*, 1–7. doi:10.1016/j.jii.2019.02.001

Zhang, R., & Chan, W. K. (2020). Evaluation of Energy Consumption in Block-Chains with Proof of Work and Proof of Stake. *Journal of Physics: Conference Series, 1584*(1), 012023. doi:10.1088/1742-6596/1584/1/012023

Zhang, X., Chen, X., & Xu, X. (2021). Smart contract-based privacy-preserving data sharing framework for blockchain. *Journal of Parallel and Distributed Computing*, *150*, 58–66. doi:10.1016/j.jpdc.2020.11.012

Zhang, Y., & Liu, X. (2019). Smart contract-based secure data sharing for cloud storage. *Journal of Cloud Computing (Heidelberg, Germany)*, *8*(1), 1–14. doi:10.118613677-019-0124-4

Zhang, Y., Sun, Y., Chen, J., & Liu, J. (2019). GasReducer: Detecting and Mitigating Gas-Costly Patterns in Ethereum Smart Contracts. *IEEE Transactions on Dependable and Secure Computing*, *17*(2), 405–419. doi:10.1109/TDSC.2018.2863135

Chapter 5
Blockchain Technology and Additive Manufacturing

Yesim Can Saglam
https://orcid.org/0000-0003-0528-6135
Gebze Technical University, Turkey

ABSTRACT

In recent years, the integration of the internet of things and information technology has become the critical technological and advancement tendency to remodel the manufacturing sectors. As a novel and advanced manufacturing technology, additive manufacturing has exhibited its splendid competency to generate products with complex geometry with the help of new materials. It is a class of manufacturing technologies that enables businesses to render innovative products and services. On the other hand, blockchain is one of the elements of the continuing digital transformation that has the potential to bring on modifications to business processes and is gaining momentum recently. Although blockchain attracts a lot of attention, few studies are available on its relations to additive manufacturing. This chapter investigates the impact of blockchain on additive manufacturing.

INTRODUCTION

From the age of steam power to the age of artificial intelligence and smart factories, many achievements have been obtained with the help of the industrial reformation. For the first time, the notion of the fourth industrial revolution, namely Industry 4.0, was offered in Germany in 2011. Since that time, many countries and manufacturing industries have embraced Industry 4.0 gaining more and more attention for enhancing their position and keeping living in a competitive environment (Shah et al., 2022).

DOI: 10.4018/978-1-6684-8639-9.ch005

Especially, manufacturing and industrial processes have tremendously evolved in a new business era thanks to Industry 4.0 facilities. It has accelerated the use of novel technologies like the robotics and internet of things. Among these technologies, Blockchain is an important technology with many advantages such as security, transparency, and decentralization (Ghimire et al., 2022:2170). Oh and Shong (2017, p.336) defined Blockchain as a "technology to secure integrity and reliability of transaction records without trusted 3rd service provider, by having all the participants in the network create, record, store and verify transaction information jointly, and has the structure to realize various application services based on distributed network infrastructure using security technologies including Hash, Digital Signature and Cryptography".

In the beginning, Blockchain technology was utilized in the finance sector (Oudani et al., 2022). In 2009, Blockchain technology was initially performed as the underlying platform premeditated to unfasten the "double-spending" issue for Bitcoin (in other words, how to transmit digital value absent of resting on a confident third aspect) (George et al., 2019). In the course of time, many supply chains and manufacturing industries have realized the advantages of this novel technology. Blockchain applies the techniques of data structure to collect as well as decode the data for whole supply chain processes constituted by customers, manufacturers, and suppliers, which will possibly culminate in structural alternations in whole kinds of the supply chain. Except that, the advancement of Blockchain functionalities in various fields of the supply chain like the logistics and distribution, transportation, security improvement, education, electric power industry, emissions trading, smart-contract transactions, governance, healthcare systems, and business information is still necessitated to more investigation (Oudani et al., 2022). Blockchain can be labelled as a technology of distributed ledger, planned in the way that network attendees enroll as well as affirm the transaction data (Oh and Shong, 2017).

Additive manufacturing which comprises many manufacturing techniques and technologies is a forceful manufacturing method and guides a paradigm shift in manufacturing. It has a significant impact on businesses, particularly in terms of the management of supply chain process (Turkcan et al., 2022). Additive manufacturing offers many impressive benefits over other manufacturing methods. Inside of them, the most significant is probably the competency of production of geometrical shapes which are challenging or in some situations infeasible to manufacture in any other manner (Butt, 2020). As a matter of principle, it presents much support for the supply chain management performance since it enhances swift innovation as well as product amendments with quick changes in design. Additive manufacturing is applied in various sectors, comprising aerospace, automotive, machinery, medical products, and electronics (Delic et al., 2019). With the development of stereolithography in the 1980s, the introduction of additive manufacturing commenced. SLA-1 by 3D Systems was the first additive manufacturing system. The development of additive manufacturing has been facilitated by technologies of desktop computers and industrial lasers (Thomas, 2016). The technology of additive manufacturing provides rapid manufacturing proportions compared to traditional investment casting as well as better accuracy (Kulkarni et al., 2021)

Despite it has various benefits, some challenges are faced in the administering process of additive manufacturing. For instance, the surface finish of the products as a result of the additive manufacturing process is weak as compared to traditional operation processes. All materials as well as colors cannot be printed, and investment is high (Kulkarni et al., 2021). In the literature, various studies call for research on additive manufacturing and especially the challenges related to its digital nature. To cope with these challenges can unlock the strengths of additive manufacturing as well as effect supply chains across many sectors (Kurpjuweit et al., 2021). In addition, various issues have been growing in significance in complex supply chains, comprising protection of intellectual property, confidence and communication between supply chain members, and maintaining visibility as well as communication and control. Experts envisage the need to apply convenient technology to render an uninterrupted exchange of information across the supply chain (Fosso Wamba & Guthrie, 2020).

The application of Blockchain technology in additive manufacturing might provide a more effective, secure, and credible additive manufacturing process. This facilitates the integration of Blockchain technology and additive manufacturing. Additive processes do have not an only, seamless data exchange that stays straight from the first production to the last output, which is why additive manufacturing is yet to be approved in a wide-ranging. Blockchain may be applied as a backend reliance layer that registers as well as affirms each incident that occurs in the additive manufacturing life cycle (Ghimire et al., 2022). In addition, while prior investigations have made important contributions to the comprehension of how Blockchain can impact supply chains on an overall stage, there is a gap in the literature that investigates the opportunities of Blockchain to overcome additive manufacturing challenges (Kurpjuweit et al., 2021).

Therefore, this study discusses the role of Blockchain in additive manufacturing. First, the concept of additive manufacturing has been explained, then it has been expanded the concept of Blockchain, and finally, it has been discussed the effects of Blockchain on additive manufacturing.

BACKGROUND

Blockchain

Ahmed et al. (2022) defined Blockchain as a "decentralized peer-to-peer network consisting of a chronological chain of blocks that are tamper-proof and cryptographically protected". The technology of Blockchain is a diversified database of archives or common private or public ledgers of whole digital incidents that have been conducted and shared between Blockchain attendees. Its history can be based on distributed ledger technology (Saberi et al., 2019). The fundamental feature of a Blockchain is the effectuation of a stable as well as a shared digital record of the time-stamped process (Ahmed et al., 2022). It is a technology that has the features of negotiation, security, automation, transparency, and immutability. It provides the attendees to have ownership and self-administrate of the data they constitute. Hence, it is entitled as "trust-free system" where no third-party affiliation is required (Dursun et al., 2022). Figure 1 demonstrates the various features of Blockchain (Ghimire et al., 2022).

Figure 1. The characteristics of blockchain technology

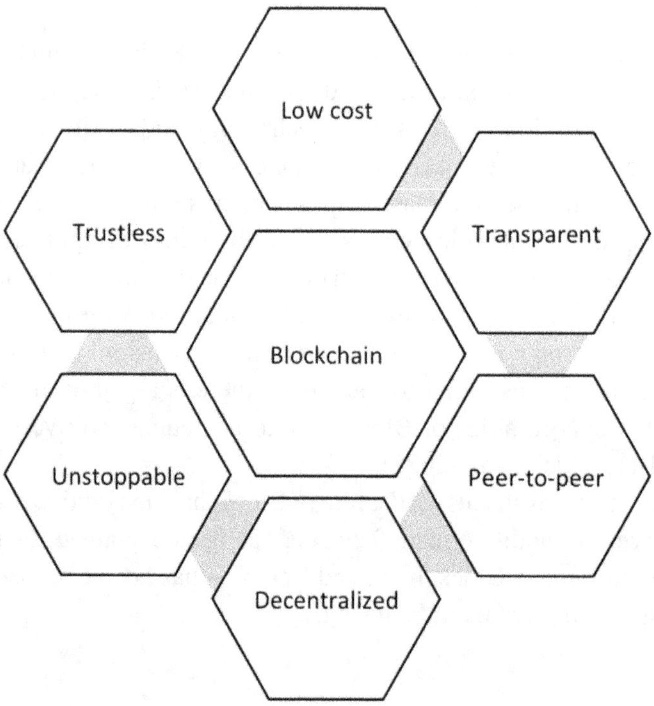

The fundamental features of Blockchain can be categorized into three titles (Gaur, 2020):

1. **Abide by Commerce, Confidence, and Ownership:** While commerce and ownership express the flow and the transfer of ledger entries, confidence expresses the unreliable nature of a transaction system. These three ingredients, commerce, confidence, and ownership, are basic to any Blockchain system.
2. **Implementation Is Basically Transactional in Nature:** It can generally be debated why the perceived benefits of Blockchain from a distributed database or a relational database cannot be achieved. It is a multilateral transaction that makes an implementation convenient for Blockchain. In particular, appropriateness necessitates long-term processes with various micro-transactions that are affirmed and legitimated through the Blockchain-powered transaction system, while databases might still be applied for progression or replication to suit the business systems.
3. **Enterprise Networks Which Constituted Non-Monopolistic Attendees:** This constitutively handles the distributed and decentralized computation models.

From a general point of view, Blockchain is in its initial phase of development (Ramachandran et al., 2022; Bag et al., 2022). Some researchers (Cunha et al., 2021) suggest that Blockchain has transformed via four primary revolutions:

- **Blockchain 1.0–Cryptocurrencies:** Blockchain 1.0 was created in 2009 and devoted to overcoming the double-spending difficulties by entering cryptocurrencies similar to Bitcoin. Before long, the Bitcoin launch, and other cryptocurrencies (option to change) evolved and connected this initial revolution.
- **Blockchain 2.0–Smart Contracts:** 2015 pronounced the commencement of a novel epoch in Blockchain with the creation of smart contracts and the notion of overall-aim programmable Blockchain, of which Ethereum is a primary illustration.
- **Blockchain 3.0–Decentralized Applications:** In 2017, a novel creation of Blockchain implementations was released. Dapps have their backend code functioning on a Blockchain (a decentralized network of computers) where warehousing, as well as transmissions, can in addition be decentralized.
- **Blockchain 4.0–Interworking and Scalability:** Since 2019, substantial studies have been done to(a) strengthen scalability, (b) expand interworking between various Blockchain, (c) merge Blockchain implementations with present enterprise resolution as well as information technology infrastructures, (d) incorporate Blockchain technology with other leading technologies (such as Internet of Things, artificial intelligence). All these requests to diffused and hasten up the acceptance of Blockchain technology.

Blockchain has been applied to massive applications in the financial sector and non-financial sectors. It has ushered in a new age in the area of information technology. Different sectors (like medical sector, financial sector, and supply chain) have applied Blockchain technology (Ghimire et al., 2022). Due to the advantages like data security, enhanced visibility, smart contracts, and enhanced supply chain management, Blockchain is gaining popularity in recent years (Bag et al., 2022). Blockchain technology can promptly effect competitive advantage by enhancing seamless information flows as well as productivity, enhancing the effectiveness of joint decision-making in the supply chain, and developing relationships with the members of supply chains and clients. Blockchain technology might improve competitive advantage and competitive performance (Fosso Wamba and Guthrie, 2020). In the last period, Blockchain can be performed in additive manufacturing (Ghimire et al., 2022). However, since it is a new application, early adopters have encountered particular challenges in performing Blockchain. Directing academics and technical professionals to question the advantages of this technology is also one of the challenges encountered by these adopters. Disruptive innovations lead to various issues as well as challenges in any market, comprising legislative, technical, social, and adoption-associated topics (Ramachandran et al., 2022).

Additive Manufacturing

Additive manufacturing (AM), generally known as 3D printing, is a means for transforming a 3D model into a physical thing. It assists in the effectuation of the accurate potential of digitization of manufacturing - design once, produce anytime, anywhere, on-demand (Haridas et al., 2022). Additive manufacturing describes a cumulative of manufacturing technologies that produce products in a layer-by-layer road, which submits an excellent facility to develop the flexibility of design and diminish the buy-to-fly percentage. This road facilitates the use of additive manufacturing products in different mission-significant industries, such as healthcare and aerospace (Shi et al., 2021). It does not necessitate tools; it administrates the production of sophisticated external shapes absent of an outstanding ascent in expenditures as well as time; it enables the production of sophisticated internal shapes at various scales (mesoscale and microscale); the shape of the last output is analogous to the core digital design. Hence, the process can be directly connected to design and manufacturing, diminishing the number of consequential processes. Throughout the production of products with this technology, the requirement of technical employees is minimal (Lopez Taborda et al., 2021).

Additive Manufacturing is a digital procedure facilitating the swift manufacturing of three-dimensional sections through layering and is recently finding prevalent application in a plethora of scientific fields (Keefe et al., 2022). It has distinctive features from traditional manufacturing (Lopez Taborda et al., 2021; Turkcan et al., 2022) and the core distinctive feature is that manufacturing is realized through abrogating material parts from solid material in traditional manufacturing, while manufacturing is performed through annexing materials layer by layer in additive manufacturing. It can transform a "3D computer-aided-design model" into a "3D physical object" through annexing material layer by layer. In consequence, additive manufacturing renders various advantages as contrasted with traditional manufacturing (Turkcan et al., 2022). Both the benefits and limitations and the reason that manufacturing is powerfully associated with the design more than with other processes are one of the core reasons that motivate the constitution of design methodologies for additive manufacturing to benefit from it and abstain from the limitations and disadvantages (Lopez Taborda et al., 2021). By means of additive manufacturing total delivery time can be decreased and additional costs of tool derangement as well as diversity are substantially lower than in conventional manufacturing (Delic et al., 2019). While the novel technology postulates to alter the supply chain to a demand chain, alluding to how clients become the ingredients of fabrication, additive manufacturing also extinguishes particular activities across the supply chain. This situation makes it attention-grabbing to comprehend those facilities that come into play as such disruption happens (Öberg, 2019). In substance, many potential benefits of additive manufacturing have been clarified in the literature as follows (Kulkarni et al., 2021):

- Remove tooling, diminishing expenses as well as the tool setup time
- Practicability of manufacturing in small batch sizes
- Adaptedness of product function optimization
- Design flexibility
- Decreased lead times
- High level of product customization
- A decrease of production waste
- Shorter supply chains
- Low inventories

The technologies of Additive Manufacturing can promptly generate parts by the residual addition of material layers applying data from three-dimensional (3D) computer models, utilizing heat or light to constitute things absent of the losses elemental with tooling, thereby, rendering important possibilities for manufacturing (Delic et al., 2019). Additive manufacturing is a fundamental technologies of Industry 4.0 as well as it is also anticipated to have a favorable effect on the production of organizations. In addition, it ensures a high level of customer involvement in designing and generating the last outputs within a manufacturing environment (Kulkarni et al., 2021). Over the past 30 years, considerable progress exists in the enhancement of additive manufacturing technologies and nowadays it provides the advantages to revolutionizing manufacturing operations and their supply chains (Delic et al., 2019). According to Klahn et al. (2020), the key characteristics of additive manufacturing can be divided into two categories:

1. The variable expenditures of additive manufacturing are nearly unconnected from the interaction of the part. Engineers might name after "complexity for free" to optimize the serviceableness of a tool as well as enhance the manufacturing process or performance of a product.
2. The constant expenditures of additive manufacturing are low since the process does not necessitate any part-particular investments in tooling programming. There is not much distinction in sum expenditures among the manufacturing of a batch of similar parts and the identical number of particular parts having an analog volume. This facility enables manufacturing at a lot size of one economically plausible.

THE ADVANTAGES OF BLOCKCHAIN TECHNOLOGY IN ADDITIVE MANUFACTURING IMPLEMENTATION

The progress of Blockchain technology is not unconnected. Reciprocal defragmentation is a significant necessity for durable developing technologies in after days, presenting those outputs, rather than being substitute goods, integrative and improving each other to cope with individual output deficiencies. Blockchain is a point of aggregation for different technologies, comprising internet of things, big data analysis, data visualization, and cloud computing (Saberi et al., 2019). These technologies have many advantages. For instance, the internet of things diminishes the expenditures of gathering colossal amounts of information, the improvement of 5G can ease the stable as well as high-speed transmission of data, and artificial intelligence enhances the intelligent analysis of big data. Therefore, Blockchain describes that colossal amounts of information could be obtained at a lower cost. In addition, network safeness as well as the value of data may be warranted, which can effectuate the rotation of the information network into the worth network as well as provide the worth of network and information entities (Li and Chen, 2022). Furthermore, Blockchain also renders a direct road for evaluating product quality during transportation. For example, by analyzing data on the travel path, members of a supply chain can be informed whether the product was in the wrong environment or whether it remained in a place for too long (Kshetri, 2018). In addition, various techniques like machine learning, machine vision, and deep learning are generally been used in the additive manufacturing process for estimating errors, deficiencies, and faults as well as enhancing the whole system (Parvanda and Kala, 2022).

Blockchain enables participating parties that do not have confidence in each other to settle over a log of transactions and other digital incidents utilizing distributed arrangement mechanisms. This extinguishes the role of a confident third side (for instance government for finances or a bank) which had conventionally solved the trust topic (Haridas et al., 2022). In a general manner, the advantages of Blockchain in additive manufacturing can be explained for three reasons (Ghimire et al., 2022):

- IP safety against public attainment to the characteristics of additive manufacturing.
- Pursuing the 3D prints through cloud platforms.
- Monetization of additive manufacturing for organizations.

In this study, the opportunities for five Blockchain features in additive manufacturing have been clarified. Table 1 presented five Blockchain attributes selected from previous studies (Ghimire et al., 2022; Ramachandran et al., 2022; Li and Chen, 2022).

Trustless

Trustless in Blockchain has particular meanings. First of all, the rubrics of Blockchain are implemented by computer codes. Computer codes, unless any obscurity in their sense, are not subject to human judgment or disturbance. Next, Blockchain is censorship-resistant. Any trading that pursues predetermined rules and provides

Table 1. The opportunities for five Blockchain properties in additive manufacturing

Blockchain Attributes	Use in Additive Manufacturing	References
Trustless	Unauthorized access to data has been diminished	Ghimire et al. (2022), Ramachandran et al. (2022)
Security	Protects data administration in additive manufacturing supply chain	Ghimire et al. (2022)
Sustainability	Collaborative governance	Li and Chen (2022); Bhubalan et al. (2022)
Information sharing	Knowledge and information have been shared	Li and Chen (2022)
Decentralized	-Enhances effectiveness and reliability for additive manufacturing chains. -In the supply chain, assists in conserving and administering activity of additive manufacturing.	Ghimire et al. (2022), Ramachandran et al. (2022)

sufficient charges is ensured to be modified by validators or miners and merged into the distributed ledger. Ultimately, there is no arrangement jeopardy due to the fact that the notification of the token explanation and the affirmation of token trading happen concurrently. Lastly, information in the scattered ledger is obvious to all and tamper-proof (Xu & Zou, 2021).

In a dynamic environment with the availability of malevolent vehicles, inaccurate as well as false messages might be sent, thereby endangering critical security problems. For instance, a false traffic update message can be shared which declares that a roadway is normal, but it is actually under construction by the malicious vehicle. Chauffeurs who read the wrong message might not pay enough attention to the roadway circumstance and endanger traffic stops very probably. Hence, it is important to guarantee that vehicles will upload valid and correct information (Wang et al., 2022). However, because of the features of additive manufacturing (such as being a new technology or having a high-speed moving vehicle), the communication relationships between vehicles are temporary and unreliable, endangering the lack of trustworthiness between vehicles and the management center. In this situation, Blockchain technology can be applied as a solution.

Blockchain has various characteristics like decentralization, immutability, anonymity, and traceability, which are convenient to scatter trust management in vehicles. Because of its anonymity, it provides vehicles to upload messages anonymously, which efficiently provides privacy as well as security of vehicles. In addition, its tamper-evident nature efficiently prohibits hackers from tampering with the content of uploaded data. When a malevolent vehicle uploads wrong knowledge, the traceability features of Blockchain can pursue the correct identity of the vehicle, which significantly diminishes the likelihood of a vehicle uploading wrong messages and solves the problem (Wang et al., 2022). Thereby, Blockchain can take away the concept of confidence from centralized businesses to a decentralized platform that is statistically affirmable as well as cryptographically secure (Prakash et al., 2022). Blockchain trust-free scattered network of nodes as well as tamper endurance properties may enhance to support client confidence. The near-zero likelihood to change formerly published blocks absent of being realized, the digital cryptographic trace of each published knowledge in the ledger, and the transparency supplied through the pursuit of knowledge published in a decentralized chain of blocks put jointly are what give rise to the Blockchain as a trust-free technology. Transparency is also associated with trust-building, since apparent conservation measures of consumers' data, labor conditions, production transparency and social responsibility consolidate trust (da Silva & Moro, 2021).

Security

The main topic associated with the products produced through additive manufacturing is the protection of intellectual property (IP) rights. Additive manufacturing is described as a peerless kind of decentralized piracy where counterfeit products can be quickly produced. The owners of 3D printers may easily purchase licensed patterns for 3D printing. However, there is an opportunity that owners would print outputs conserved through IP rights absent of attaining the required rights (Sekerin et al., 2021). In other words, the preservation of IP has been a critical issue as smart factories developed in additive manufacturing. The significant properties of Blockchain are that it is scattered, affirmed, and constant. Scattered architecture describes that systems do not subject to any centralized power yet utilizes a peer-to-peer trawl of computer servers retained by decentralized interest proprietary. Every proprietary deposits a copy of the entire Blockchain. Nonetheless, Blockchain might pursue safeness as each transaction is affirmed through utilizing public-private-key cryptography as well as the commerce records on the blocks might not be regulated when they are acknowledged as ingredients of the chain since they are glued to each other. Any interference with a transaction record can be informed by many computers in the network. Thus and so, the uniformity of data substance is affected (Helo & Hao, 2019).

Various features of Blockchain technology such as distributed ledger, smart contracts and decentralization are associated with IP protection (Haridas et al., 2022). In addition, the process of additive manufacturing is missing a seamless and single data strand that leads accurate from the first generation to the final outputs. For this reason, on a larger scale, additive manufacturing has yet to be adopted. Blockchain can be applied as the safeness layer that safes as well as records incidents in the life cycle of additive manufacturing (Ghimire et al., 2022). Just like in logistics, applying Blockchain technology in additive manufacturing can substantially decrease the need for the affirmation of paper documents, besides it can enhance the safety and efficiency of information sharing about the process of additive manufacturing (Li & Chen, 2022). According to research by Can Saglam et al. (2020), the primary factor is the security of information for managers when sharing risk-related issues in the supply chain, particularly during the risk mitigation activities, which has strengthened these necessities to restrain the further circulate of the risk. Additive manufacturing requires specialized knowledge and can be easily imitated by competitors. Therefore, the information about the production process of additive manufacturing should be confidential. When thinking about this situation in additive manufacturing, Blockchain technology can support the information visibility of information in the whole additive manufacturing process and it is hard to blend with Blockchain data. When Blockchain technology has been used in additive manufacturing the information about all phases of additive manufacturing is present to whole partners in the supply chain, which provides the safeness of process information.

Sustainability

Constantly increasing environmental pollution, swift exhaustion of resources, and global warming have diminished biological diversity which leads to deterioration in the ecological balance. this situation leads companies to take precautions in environmental matters. Because businesses are generally seen as the source of environmental problems. So, businesses have had to revise their production processes as a consequence of pressure from the governments as well as the community (Yildiz Çankaya & Sezen, 2019). Recently, many businesses have endeavored to related services or products that were not associated with the prior to generate a dynamic and networked ecosystem through the mobile internet, artificial intelligence, and other different technologies. From client request to commodity supply, the supply chain is altering from one-way administration to a collaborative management function. Accordingly, it is critical to comprehend the operation administration of businesses from the angles of the ecosystem (Li & Chen, 2022). Sustainability can be defined as "meeting the needs of the current generations without compromising the ability of future generations to meet their own needs". This description has highlighted three significant dimensions of sustainability: social, environmental, and economic (Çankaya & Sezen, 2020). Social sustainability defines as "ensuring efficient use of natural resources today and in the future through developing and preserving the social environment that will support human needs and environmental sustainability" (Buckingham-Hatfield & Evans, 1996). Social sustainability is associated with the responsibility of the enterprises for the overall community. Environmental sustainability is associated with the environmental effect of business activities. The origins of environmental problems (such as supply, production, and product) require to be explained for enhancing a better comprehension of the significant environmental problems as well as create efficient solutions. Economic sustainability is enhanced when enterprises do not endanger any deterioration in the natural environment while pursuing their purpose of profit. Consequently, the core of sustainable improvement is the economic side of the firms which constitutes profit, generates jobs, and develops social welfare (Çankaya & Sezen, 2020).

In a sustainable operation aspect, energy consumption is a critical topic to be focused. Additive manufacturing provides an energy-efficient manufacturing approach, especially in the aspect of waste characteristics and the environment. It is evaluated that additive manufacturing has a great scope for material waste and diminishes energy consumption during the manufacturing process (Ramesh &

Vinodh, 2021). According to studies, additive manufacturing has three core benefits over conventional manufacturing. Firstly, this manufacturing process decreases the restraints of conventional manufacturing and enhances the freedom of design for innovative processes and products. Next, additive manufacturing benefits like diminished transport distances as well as inventories provide businesses to augment their margin. Lastly, additive manufacturing has significant potential to diminish the environmental effects of conventional manufacturing (Çankaya & Sezen, 2020).

A critical implementation of Blockchain is probably to be in affirming sustainability. Applying Blockchain, it is possible to make indicators associated with sustainability more measurable and more meaningful. Hence, Blockchain has the potential to end unethical and illegal implementations as well as practices (Kshetri, 2018). The relationship of businesses has been altered from a competitive to a Blockchain technology and symbiotic relationship can ensure novel management methods for collaboration in the supply chain. For instance, in 2020, the Ministry of Ecology and Environment transmitted "Several Opinions on Promoting the Modernisation of the Ecological Environment Monitoring System and Monitoring Capability (Draft for Comment)", which propounded developing the ecological environment pursuing technology system, improving smart pursing as well as enhancing the implementation of the Blockchain, internet of things, artificial intelligence, sensors, and other novel technologies. Technology may provide to monitor the carbon emissions of the whole process of the supply chain and assist in pursuing the real-time information of the overall ecological environment following the system, which can diminish discrepancies among various stakeholders on carbon emissions and, thereby, assist to the collaboration of various stakeholders among carbon trading systems (Li & Chen, 2022). Besides, Blockchain can promote sustainability in multiple ways through information transparency, automation, and reliability which can efficiently enhance sustainability initiatives. In addition, information about green products like the source of material, processes, involved actors, and other related information can be made present on the ledger. Furthermore, Blockchain technology may help in achieving sustainability aims because of its distributed ledger technologies features in conjunction with the functionality of smart contracts and tokenization. Redesigning plastics products benefiting molecular tagging which comprises both Blockchain technology and sustainability would provide to better solutions for plastic pollution in the not too distant future (Bhubalan et al., 2022).

Information Sharing

Information sharing in additive manufacturing has critical importance because the process necessitates specialist information. However, in the supply chain many obstacles which hinder sharing of information exist. This can be related to technical or security issues (Can Saglam et al.., 2020). With using of Blockchain technology facilities, information-sharing facilities can be obtained more and the barriers to information-sharing can be overcome. For instance, a social media application relying on Blockchain technology called as Sola eases the sharing as well as independent interaction of information (Li & Chen, 2022).

Blockchain-related to an environment to application contracts in a non-centered and secure manner enables an essential platform to introduce automation to present and validate these contracts. While the constitution or monitoring stage of the process, Blockchain can be applied to record all relevant information regards to the side being printed, comprising the design used, calibration of the machines used, etc. (Haridas et al., 2022). Especially, the afflux of information among the supply chains has been developed in the previous three decades by the improvement as well as readjustment of information systems, Blockchain in supply chain administration proceeding improvement of cloud-based enterprise software as well as communication technologies (Ahmed et al., 2022). Implementation of Blockchain can save costs as well as enhance performance by abstaining human error as well as paper-based procedures with the help of an efficient information-sharing process (Ramachandran et al., 2022).

Decentralized

The notion of decentralization in business has grown by leaps and it has already found its scope in supply chain management (Prakash et al., 2022). A specific property of operating in a digital environment is the likelihood of generating programs and algorithms that may be comparatively or thoroughly performed or enforced when particular situations happen without human inferences—a property generally known as "smart contracts" (Francisco and Swanson, 2018). Contrary to traditional production, additive manufacturing has a decentralized production system by nature. The decentralized nature of additive manufacturing provides flexible production for businesses operation (Sekerin et al., 2021). Blockchain technology supports and enhances this feature of additive manufacturing. One of the initial likely functionalities is to implement Blockchain to pursue whole implementations in the supply chain. For instance, who is operating the actions, where the location of each action is, and at what time. Each member of the supply chain may pursue shipments, outputs, deliveries, and progress. In addition, they can lightly evaluate the performance of every stage in the supply chain and pursue the quality of outputs along transportation. Hence, a Blockchain-based supply chain diminishes the workload as well as provides traceability, while augmenting effectiveness, diminishing expenditures, and warranting more reliance that the outputs are authentic as well as of high quality (Helo & Hao, 2019).

The implementations and usability of Blockchain technology in supply chains are ascending incessantly with the advent of the machines as well as Internet of things rendering operational data directly. It is more impressive to pursue and control things from their roots by applying the Internet of things. Moreover, traceability is blocked when material information is fragmental or missing. Nevertheless, the benefits of traceability are limited by the interaction in the supply network (Francisco and Swanson, 2018). Blockchain creates a secure environment in the decentralized production system, providing information sharing and thus increasing traceability. Blockchain augments the decentralization of the database which eventually increases the security of each transaction. This decentralization administrated through more than one user is generally known as a piece of distributed ledger technology. The features of distributed ledger technology are distributed, time-stamped, immutable, unanimous, secure, and programmable. Hence, the Blockchain is a part-to-part distributed ledger that is cryptographically append-only, secure, immutable (very difficult to alter), and updateable solely by agreement among partners (Bhubalan et al., 2022).

IMPLICATIONS FOR MANAGERS

Governments and businesses are actively enhancing innovation as well as the value of Blockchain in the supply chain. In the area of supply chain management, Blockchain has critical potential for wide implementation. In addition, it has attracted significant attention in both the industrial circles and academic. The integration of Blockchain technology in additive manufacturing is gaining importance. The enhancement of emerging businesses and the burgeoning of multinational businesses are associated with Blockchain technology and additive manufacturing. As Blockchain can develop the collaboration and innovation of a business by unifying a diversity of resources via a decentralized platform, it has an effect on both business profits and administration models. Enterprises should comprehend the role of Blockchain in the additive manufacturing process if they endeavor to develop the efficiency of the business process.

The critical remarks of the literature for managers can be presented as follows (Ghimire et al., 2022):

- The intellectual property which is an intangible asset of the firm which encircled the additive manufacturing process can be consolidated, licensed, and secured with the help of the Blockchain. Applying Blockchain technology to assure Copyright issues generates a ‘‘Chain of Trust.” For instance, an individual X1 can permit another individual X2 to print out five copies of a specific output. The licensing knowledge is reserved in the Blockchain through an intelligent contract that solely provides X1 and X2 to reach it.

- By establishing an information-sharing mechanism, additive manufacturing and Blockchain may effectuate more personalization as well as a customer-centered attitude to the operation duration. Customers of the businesses must be fully comprised in the output lifecycle, which from the design to the finished product.
- As a particular fragment of Industry 4.0, cyber-physical systems as well as technologies like the internet of things are being applied to additive manufacturing. for storing and analyzing real-world data and information, these cyber-physical systems are beneficial. ın additive manufacturing, Blockchain technology can be applied to ensure an assurance network by which the internet of things might be authenticated.
- Last but not least, data can be obtained securely from clients to analyze further optimize and develop services. Blockchain can render stakeholders the impression of trust and transparency in the chain.

CONCLUSION

The effect of the fourth industrial revolution, commonly known as Industry 4.0, cannot be abnegated since it has sparked off both academic and industrial research activities. The utilization of novel technologies and enhance of the workforce are critical to surviving in a highly competitive environment. The integration of digital and tangible worlds has alternated operation practices as well as has nestled many advantages that were unimaginable previously (Butt, 2020). With the implementation of Blockchain and additive manufacturing, it has been possible. Nevertheless, there are still some topics that should be clarified. In addition to the development of Blockchain and additive manufacturing, other aspects which elucidate the integration of this topic also require attention. Therefore, this chapter focus on the role of Blockchain technology in additive manufacturing.

This chapter indicates that, in the matter of information sharing, establishing trust and security, enhancing sustainability and decentralized, Blockchain technology encounters the necessaries of whole additive manufacturing for the authenticity of transaction data. In the field of additive manufacturing investigations on the Blockchain is still in its infancy and should contemplate probable applications. Several businesses declare that they have very little information related to Blockchain and that they evaluate the effect of Blockchain as a jeopardy (Helo & Hao, 2019). To obtain their own first-hand experience, enterprises must conduct low-end experiments.

REFERENCES

Ahmed, W. A. H., MacCarthy, B. L., & Treiblmaier, H. (2022). Why, where and how are organizations using blockchain in their supply chains? Motivations, application areas and contingency factors. *International Journal of Operations & Production Management*, *42*(12), 1995–2028. doi:10.1108/IJOPM-12-2021-0805

Bag, S., Rahman, M.S., Gupta, S., & Wood, L.C. (2022). Understanding and predicting the determinants of blockchain technology adoption and SMEs' performance. *The International Journal of Logistics Management*. doi:10.1108/IJLM-01-2022-0017

Bhubalan, K., Tamothran, A. M., Kee, S. H., Foong, S. Y., Lam, S. S., Ganeson, K., Vigneswari, S., Amirul, A.-A., & Ramakrishna, S. (2022). Leveraging blockchain concepts as watermarkers of plastics for sustainable waste management in progressing circular economy. *Environmental Research*, *213*, 113631. doi:10.1016/j.envres.2022.113631 PMID:35714685

Buckingham-Hatfield, S., & Evans, B. (1996). Achieving sustainability through environmental planning. In S. Buckingham-Hatfield & B. Evans (Eds.), Environmental Planning and Sustainability. John Wiley & Sons.

Butt, J. (2020). Exploring the interrelationship between additive manufacturing and Industry 4.0. *Designs*, *4*(2), 13. doi:10.3390/designs4020013

Can Saglam, Y., Sezen, B., & Çankaya, S. Y. (2020). The inhibitors of risk information sharing in the supply chain: A multiple case study in Turkey. *Journal of Contingencies and Crisis Management*, *28*(1), 19–29. doi:10.1111/1468-5973.12285

Çankaya, S. Y., & Sezen, B. (2020). Industry 4.0 and Sustainability. In Handbook of Research on Creating Sustainable Value in the Global Economy (pp. 67-84). IGI Global.

Cunha, P. R. D., Soja, P., & Themistocleous, M. (2021). Blockchain for development: A guiding framework. *Information Technology for Development*, *27*(3), 417–438. doi:10.1080/02681102.2021.1935453

da Silva, C. F., & Moro, S. (2021). Blockchain technology as an enabler of consumer trust: A text mining literature analysis. *Telematics and Informatics*, *60*, 101593. doi:10.1016/j.tele.2021.101593

Delic, M., Eyers, D. R., & Mikulic, J. (2019). Additive manufacturing: Empirical evidence for supply chain integration and performance from the automotive industry. *Supply Chain Management*, *24*(5), 604–621. doi:10.1108/SCM-12-2017-0406

Dursun, E., Ulker, Y., & Gunalay, Y. (2022). Blockchain's potential for waste management in textile industry. Management of Environmental Quality. doi:10.1108/MEQ-03-2022-0085

Fosso Wamba, S., & Guthrie, C. (2020). The impact of blockchain adoption on competitive performance: The mediating role of process and relational innovation. *Logistique & Management, 28*(1), 88–96. doi:10.1080/12507970.2019.1679046

Gaur, N. (2020). Blockchain challenges in adoption. *Managerial Finance, 46*(6), 849–858. doi:10.1108/MF-07-2019-0328

George, R. P., Peterson, B. L., Yaros, O., Beam, D. L., Dibbell, J. M., & Moore, R. C. (2019). Blockchain for business. *Journal of Investment Compliance, 20*(1), 17–21. doi:10.1108/JOIC-01-2019-0001

Ghimire, T., Joshi, A., Sen, S., Kapruan, C., Chadha, U., & Selvaraj, S. K. (2022). Blockchain in additive manufacturing processes: Recent trends & its future possibilities. *Materials Today: Proceedings, 50*, 2170–2180. doi:10.1016/j.matpr.2021.09.444

Haridas, A., Samad, A. A., Vysakh, D., & Pathari, V. (2022, March). A blockchain-based platform for smart contracts and intellectual property protection for the additive manufacturing industry. In *2022 IEEE International Conference on Signal Processing, Informatics, Communication and Energy Systems (SPICES)* (Vol. 1, pp. 223-230). IEEE. 10.1109/SPICES52834.2022.9774219

Helo, P., & Hao, Y. (2019). Blockchains in operations and supply chains: A model and reference implementation. *Computers & Industrial Engineering, 136*, 242–251. doi:10.1016/j.cie.2019.07.023

Keefe, E. M., Thomas, J. A., Buller, G. A., & Banks, C. E. (2022). Textile additive manufacturing: An overview. *Cogent Engineering, 9*(1), 2048439. doi:10.1080/23311916.2022.2048439

Klahn, C., Fontana, F., Leutenecker-Twelsiek, B., & Meboldt, M. (2020). Mapping value clusters of additive manufacturing on design strategies to support part identification and selection. *Rapid Prototyping Journal, 26*(10), 1797–1807. doi:10.1108/RPJ-10-2019-0272

Kshetri, N. (2018). Blockchain's roles in meeting key supply chain management objectives. *International Journal of Information Management, 39*, 80–89. doi:10.1016/j.ijinfomgt.2017.12.005

Kulkarni, P., Kumar, A., Chate, G., & Dandannavar, P. (2021). Elements of additive manufacturing technology adoption in small- and medium-sized companies. *Innovation & Management Review*, *18*(4), 400–416. doi:10.1108/INMR-02-2020-0015

Kulkarni, P., Kumar, A., Chate, G., & Dandannavar, P. (2021). Elements of additive manufacturing technology adoption in small- and medium-sized companies. *Innovation & Management Review*, *18*(4), 400–416. doi:10.1108/INMR-02-2020-0015

Kurpjuweit, S., Schmidt, C. G., Klöckner, M., & Wagner, S. M. (2021). Blockchain in additive manufacturing and its impact on supply chains. *Journal of Business Logistics*, *42*(1), 46–70. doi:10.1111/jbl.12231

Li, Y., & Chen, T. (2022). Blockchain empowers supply chains: challenges, opportunities and prospects. *Nankai Business Review International*. doi:10.1108/NBRI-06-2022-0066

Lopez Taborda, L. L., Maury, H., & Pacheco, J. (2021). Design for additive manufacturing: A comprehensive review of the tendencies and limitations of methodologies. *Rapid Prototyping Journal*, *27*(5), 918–966. doi:10.1108/RPJ-11-2019-0296

Öberg, C. (2019). Additive manufacturing – digitally changing the global business landscape. *European Journal of Management and Business Economics*, *28*(2), 174–188. doi:10.1108/EJMBE-11-2018-0116

Oh, J., & Shong, I. (2017). A case study on business model innovations using Blockchain: Focusing on financial institutions. *Asia Pacific Journal of Innovation and Entrepreneurship*, *11*(3), 335–344. doi:10.1108/APJIE-12-2017-038

Oudani, M., Sebbar, A., Zkik, K., El Harraki, I., & Belhadi, A. (2022). Green block-chain based IoT for secured supply chain of hazardous materials. *Computers & Industrial Engineering*, *108814*. Advance online publication. doi:10.1016/j.cie.2022.108814

Parvanda, R., & Kala, P. (2022). Trends, opportunities, and challenges in the integration of the additive manufacturing with Industry 4.0. *Progress in Additive Manufacturing*, 1-28. doi:10.1007/s40964-022-00351-1

Prakash, R., Anoop, V. S., & Asharaf, S. (2022). Blockchain technology for cybersecurity: A text mining literature analysis. *International Journal of Information Management Data Insights*, *2*(2), 100112. doi:10.1016/j.jjimei.2022.100112

Ramachandran, R., Babu, V., & Murugesan, V. P. (2022). The role of blockchain technology in the process of decision-making in human resource management: a review and future research agenda. Business Process Management Journal. doi:10.1108/BPMJ-07-2022-0351

Ramesh, P., & Vinodh, S. (2021). Analysis of factors influencing energy consumption of material extrusion-based additive manufacturing using interpretive structural modelling. *Rapid Prototyping Journal*, *27*(7), 1363–1377. doi:10.1108/RPJ-05-2020-0100

Saberi, S., Kouhizadeh, M., Sarkis, J., & Shen, L. (2019). Blockchain technology and its relationships to sustainable supply chain management. *International Journal of Production Research*, *57*(7), 2117–2135. doi:10.1080/00207543.2018.1533261

Sekerin, V. D., Slepov, V. A., Gayduk, V. I., Bank, S. V., & Kravets, E. V. (2021). Blockchain technology development as tool for enhancing security in management and protection of intellectual property rights in additive manufacturing. *Revista Geintec-Gestao Inovacao E Tecnologias*, *11*(2), 1184-1200. doi:10.47059/revistageintec.v11i2.1747

Shah, K., Patel, N., Thakkar, J., & Patel, C. (2022). Exploring applications of blockchain technology for Industry 4.0. *Materials Today: Proceedings*, *62*(13), 7238–7242. doi:10.1016/j.matpr.2022.03.681

Shi, Z., Kan, C., Tian, W., & Liu, C. (2021). A Blockchain-based G-code protection approach for cyber-physical security in additive manufacturing. *Journal of Computing and Information Science in Engineering*, *21*(4), 041007. Advance online publication. doi:10.1115/1.4048966

Thomas, D. (2016). Costs, benefits, and adoption of additive manufacturing: A supply chain perspective. *International Journal of Advanced Manufacturing Technology*, *85*(5), 1857–1876. doi:10.100700170-015-7973-6 PMID:28747809

Turkcan, H., Imamoglu, S. Z., & Ince, H. (2022). To be more innovative and more competitive in dynamic environments: The role of additive manufacturing. *International Journal of Production Economics*, *246*, 108418. doi:10.1016/j.ijpe.2022.108418

Wang, S., Hu, Y., & Qi, G. (2022). Blockchain and deep learning based trust management for Internet of Vehicles. *Simulation Modelling Practice and Theory*, *120*, 102627. doi:10.1016/j.simpat.2022.102627

Xu, Z., & Zou, C. (2021). What can blockchain do and cannot do? *China Economic Journal*, *14*(1), 4–25. doi:10.1080/17538963.2020.1748968

Yildiz Çankaya, S., & Sezen, B. (2019). Effects of green supply chain management practices on sustainability performance. *Journal of Manufacturing Technology Management*, *30*(1), 98–121. doi:10.1108/JMTM-03-2018-0099

KEY TERMS AND DEFINITIONS

Information Technology: Information technology can be described as a utilization of computer systems or devices to obtain as well as access information and knowledge.

Intellectual Property: Intellectual property describes a generation of consciousness like inventions, artistic works, and literary, symbols, names, and images utilized in commerce.

Chapter 6
From Manual Automation to Hyperconnection:
The Evolution and Development of Organizational Processes in Industry 4.0

Luis-Eduardo Faugier-Contreras
Universidad Veracruzana, Mexico

Karla Fernanda Guevara-Flores
https://orcid.org/0009-0004-9360-5929
Universidad Veracruzana, Mexico

José-Guillermo Hernández-Calderón
https://orcid.org/0000-0002-8014-8819
Universidad Veracruzana, Mexico

ABSTRACT

The implementation of the IoT in the industry promotes the automation of production processes in addition to the reduction of operating costs. However, its implementation is not a trivial task because it is necessary to evaluate the integration potential, analyze the different processes where its implementation is viable, seeking the most significant impact, in addition to minimizing existing technological barriers such as providing computer equipment to the industry in addition to the interconnection of devices and machines without neglecting data security in this highly connected environment. The data collected by the different devices must be able to be visualized through indicators of interest, considering these environments as intelligent environments. This chapter seeks to generate a more assimilated theoretical notion about the development and implementation of the internet of things in the industry to help the adaptation of this type of technology in their production processes and services in public or private organizations.

DOI: 10.4018/978-1-6684-8639-9.ch006

INTRODUCTION

In recent years, the analysis of processes related to administrative and management sciences has taken on particular relevance for various multidisciplinary (and in some cases interdisciplinary) studies due to the growing interest in how these processes, which develop internally within an industry, and respectively but not necessarily, an organization, impact and interact with the complex and constant network of changes we are experiencing in the "digital" and "interconnected" era that we are going through.

Naturally, fulfilling a task, function, or job service is linked to a way of proceeding with its execution, that is, that which structures certain work actions in an orderly and specific way intending to complete what is produced most satisfactorily. From the above, it is understood that every organizational process focuses on making something specific, with that "something" as the primary cause for said process. However, not all procedures are linear and self-conclusive. The larger the operation, the greater the number of different techniques (connected by a series of links) that will produce this productive conclusion.

Nonetheless, it is impossible to homogenize and summarize every organizational process into a succession of beginning, development, and end based on the production it pursues, especially since, although it has been exposed that this is the medullary purpose of the existence of a process, they are also linked to social, cultural, scientific, political, and specific technological changes.

Within administrative sciences, the study of the evolution, advancement, and way of executing organizational processes can be divided into trends or labor "fads" adopted and labeled by waves, starting from industry 1.0 (mechanization as a predecessor to automation) and currently going through industry 4.0 (total automation via hyperconnectivity and IoT).

Technology, being something that has developed in parallel and inherent to human beings, has been applied in administrative, labor, and organizational dimensions to simplify, facilitate, and increase the action of working and, with this, allow to increase production than would be obtained by performing a manual process. Since the beginning and boom of devices capable of performing different types of readings and connections with each other and the Internet, the focus on structuring an organizational process has changed radically due to all the possibilities this allows. In recent years, the number of devices connected to the Internet has increased because of the trend toward incorporating connectivity mechanisms into everyday objects to transmit or receive data. These networks and devices are related to a Computer Science Area: The Internet of Things (IoT).

Nowadays, the miniaturization of computing devices and the ubiquity of their use have brought new ways to transform industries. In addition to incorporating the

IoT devices, Context-Aware information aligned with the organization's objectives is required to wisely improve and provide relevant information, notify that something is going wrong, or interact with the devices remotely. Ubiquitous Computing (UbiComp), Ambient Intelligence (AmI), and Intelligent Environments (IEs) are three different areas that perform research related to this. The IoT device's main objective in the Industry is to collect data related to and generated in various processes to increase their autonomy and assist the decision-makers in improving the products' efficiency, effectiveness, and quality. There has been growing interest in identifying Activities and Actions in this kind of system, mainly related to the well-being of the workers in different environments. For this to happen, it is essential to consider the context and daily processes in the industry environment.

The Internet of things can be incorporated into agriculture, energy systems, smart cities, smart infrastructure, and health services. However, it is also applied prominently in the industrial environment by incorporating IoT devices in industrial automation processes, giving rise to the Industrial Internet of Things or Industry 4.0. It is essential to highlight that the research agenda for this area has been focused on improving networking and communication technology, data management, security, privacy, and power consumption to minimize expenses and increase feasibility, particularly in industries in emerging countries.

The implementation of the IoT in the Industry promotes the automation of production processes in addition to the reduction of operating costs; However, its implementation is not a trivial task because it is necessary to evaluate the integration potential, analyze the different processes where its implementation is viable, seeking the most significant impact in addition to minimizing existing technological barriers such as providing computer equipment to the Industry in addition to the interconnection of devices and machines without neglecting data security in this highly connected environment. In addition, the data collected by the different devices must be able to be visualized through indicators of interest, considering these environments as intelligent environments.

While administrative sciences define their strategies and methods of application through the industrial trends they go through and, with this, apply tools to structure their processes differently, the question arises of how IoT has hybridized in such an organic way with industry 4.0. Is IoT itself the trigger for industry 4.0, or is it instead an enhancer of it? In what ways has IoT been applied in Industry 4.0, and how have they been impacted? What advantages and disadvantages has IoT brought about in process design, and how has this affected the worker, the organization, and production? Also, Considerations related to properly implementing these devices in an established company without diminishing daily activities will be addressed.

Based on a theoretical and empirical horizon, approached from an interdisciplinary perspective from computer science and administrative sciences, this chapter seeks to provide answers to the questions posed, intending to generate a more assimilated theoretical notion about the development and implementation of the Internet of Things in the Industry to help the adaptation of this type of technology in their production processes and services in public or private organizations.

EVOLUTION AND TRANSFORMATION OF ORGANIZATIONAL PROCESSES

After the health contingency caused by COVID-19, rethinking how knowledge is assimilated from the different scientific and technical perspectives of the different fields of knowledge has become a task crucial and necessary to understand the changes humanity is experiencing, making it necessary to detect the transitions between industrial revolutions that could have been accelerated or "blurred" due to the same effect that the pandemic crisis had on all the knowledge schemes developed and known by humanity (Becerra et al., 2022; Van Gigch, 2002). The economic-administrative sciences, being no stranger to these processes, played a crucial role due to how they had to reorganize themselves at economic, organizational, and labor levels, such as developing new safety rules for employees, updating techniques and forms of production, implementing new regulations as well as a very sudden modification of the development of their processes (Reynaldos-Grandón, 2020).

The dynamics of the pandemic, together with the conditions that were established around it to contain the situation as much as possible, produced, in fact, accessibility in the way in which the so-called "digitality" or automation was adopted (Barrientos-Avendaño et al., 2020). There is evidence of an increase in businesses that migrated to digital platforms, as well as others that had to manage to implement strategies that would allow them to take that step, seeking to introduce at least the option of transferring cash for payment from a smartphone, concluding in research on local industries the enormous impact that this transition had on the operation of businesses (Sánchez-Moncayo, 2021).

Although the health situation has somewhat normalized for the period 2023, it has been made clear that there was undeniably a contextually driven update by the pandemic to be able to implement digital and electronic technologies in the organizational processes of any commercial sector in organizations of any size. This includes everything from the small corner store with the option of payment by transfer, CoDi, or NFC, to the international supermarket chain with an online shopping service (García, 2021). All these process updates were not isolated developments in which economic-administrative sciences played a significant role. They were designed in concert with other sciences to streamline how any business produces something.

In this stage of innovation, everyday things are already something more than mere entertainment; they have the quality of connecting and can maintain constant (or many times uninterrupted) connectivity within a dimension that modern humans inhabit without needing to be tangibly there: the Internet (Cano-Orón & Llorca, 2018). This new kind of device developed from the perspective of constant connectivity makes up what is known as the Internet of Things (IoT) (Gomez, 2022) and in the context of a global pandemic that sought to avoid physical contact as much as possible or crowd people together for as short a time as possible generated multiple implementation ideas, one of them its application in the Industry. So, does the impact of the implementation of technology on organizational processes allow us to review the history of industrial revolutions from another perspective? The search to make organizational processes more efficient started from the administrative sciences, with their search to implement mechanisms that would make them more effective and efficient in processes (formerly merely productive). The practical assimilation of different knowledge, techniques, and technologies from other areas of knowledge that modify the production processes of any organization directed to any industrial sector, regardless of its size, needed to be modified by what the exact economic-administrative sciences describe under the name of "Industry 4.0". (Lopez-Perez, 2020).

With this renewed perspective on the relationship between technological development and industry processes, it is considered relevant to revisit the "revolutions" industries before Industry 4.0. Doing this is intended to understand better why the Industry currently behaves in this way in the sense of practical assimilation towards the IoT and, with this, to answer if the success of the implementation of organizational processes by the IoT is the cause, consequence, or product of Industry 4.0 (Coyoli-Pereyra,2022). In this order of ideas, the perspective that the following chapter adopts is to seek to understand the notion on which the industrial revolutions that have gone through have been built and the modification that this has brought to their organizational processes, thus managing to establish a logic of how to configure these processes (Clavijo-Quintero, 2022).

Suppose we follow the history of the administration. In that case, we will find that it has the particularity of being explained through the industrial revolutions it underwent, generating a technical account of why it developed this way. These technical and organizational advances clearly show how economic-administrative sciences implement technologies to simplify the process (Papadopoulos et al., 2023). However, why do efforts toward organizational processes occur this way? Unlike the traditional look that administrative and economic schools will have on the succession of events that led and marked these "revolutionary" stages regarding efforts to define a way of producing something specific from any industrial sector, the following this proposal seeks to argue that the history of industrial revolutions can be split into two stages according to the Figure 1.

Figure 1. Division of industrial periods through their revolutions
Source: Based on Coyoly-Pereyra (2022) and Ferrer (2022)

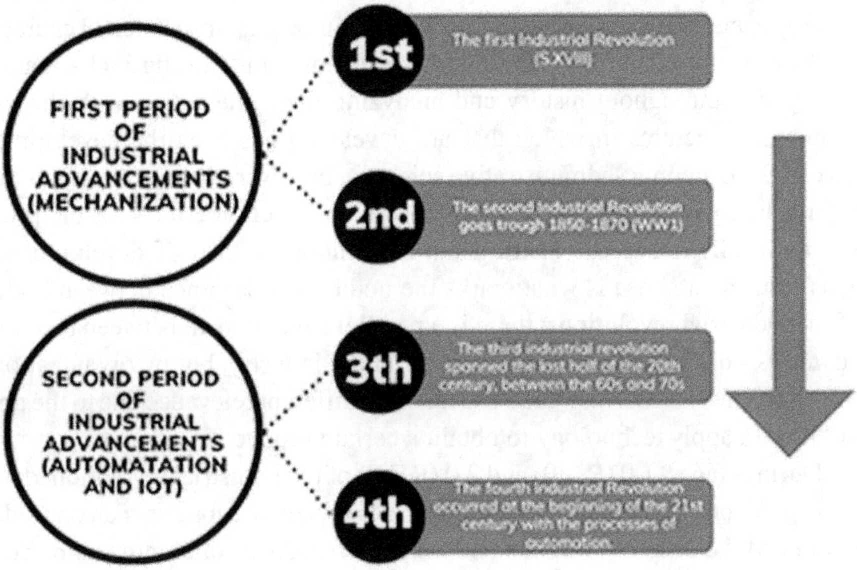

On the one hand, the first stage of the Industrial Revolution deals with establishing a production that would meet the condition of being "consumable" in any aspect (Ikbal & Khizar, 2022; Zizic et al., 2022). This is natural because initially, these efforts were founded on the private interests of mass production of any product or service launched on the market. Although the socio-historical factors that involve the motivations for this action go beyond the approaches of this writing and are widely covered by the social sciences, it is essential to clarify that the application of merely "mechanical" machinery (steam drive, simple engines, use of physics and electronic instrumentation of the early days, among other examples of technology) was not defined in terms of auxiliary, communicate, connect or facilitate. Instead, its leading development was based on the increase in production, generation, and promotion.

This previous knowledge about how the Industrial Revolutions dealt with the notion of production is essential since it exposes a substantial difference between the 1.0 and 2.0 revolutions against the 3.0 and 4.0. The technological advances that we have were given by years of development on existing technologies and the knowledge generated in the different scientific and knowledge branches that exist; this has led humanity to drive a steam engine, introduce artificial intelligence to a car, and have a device permanently connected to the Internet, that is, it is a cumulative knowledge that is generated through the research effort to develop or cover needs.

However, for the industrial theme, we find instead an instrumental character that marks its history, in effect, the way of structuring industries through organizations of any size throughout history and analyzing their interaction with the different economic structures in which that are developed has been the development and task of the economic-administrative sciences, however, the way in which the work efforts are managed and focused (that is, the production efforts for the product or service itself) rather use instruments of technology than of development itself. This instrumental use is what marks the notorious difference between both stages of the industrial revolutions; there is a parallel development between technological advances - digital with their application in the Industry, but by organizations (and the industries themselves), their use takes on different relevance due to the approach of "how to apply technology to obtain a certain product?

During stages, 1.0 (R.1.0) and 2.0 (R.2.0) of the industrial revolution, due to this mass production and high competitiveness is driven by capitalism on one side of the world and the different Marxist and initial perspectives of communism- Socialism towards self-production on the other hand, produced an enormous need around the ability to do "more of a better way in the shortest possible time." From that comes the concept of mechanization that was had since R.1.0, which, for R.2.0, becomes more evident (Medina-Valdés et al., 2022). During R.2.0, this concept of production lines assisted in organized shifts or simply the improvement of the product through the ability to reduce downtime, worker fatigue, internal accidents, and poor product quality, among other problems (Jiang et al., 2022). Before the logic previously described the actions that motivate the development of organizational processes. If R.1.0 was defined from mechanization, R.2.0 is defined from the search for mass production assisted by new energies such as electricity. Precisely, this set of factors evidenced that although to define revolutions from the economic-administrative sciences, it is applied as an instrument, the logic of its use did follow a sequence that developed and ended in the second period of industrial revolutions, partly due to the ease of accessibility presented by the growing connection capacity of the things that surround us to an interconnected network that would be known as the Internet (Ferrer, 2022).

In this second period, which can be defined as the one that encompasses Industry 3.0 and 4.0, efforts, mainly by administrative sciences, focused on continuing to implement technology that applies to their organizational processes (Blanc et al., 2021). In effect, the technological advances that existed, together with these new approaches around the idea of work that Industry 3.0 brought with it, applying knowledge from computer science and information technology, were focused on the search for automation and technological implementation.

Thus, during this period, the creation of processes focused even more on terms such as "automating," "connecting," "reducing time," "increasing," and "applying." Note that if before, strategies that fully encompassed human capabilities such as time and motion management, production lines per workstation, and attending small businesses by one person, and all those that could exist within any industry, were sought, now, in Industry 3.0, automation of specific tasks is sought in order to develop and reorganize organizational processes within an organization (Laffi & Boschma, 2022).

During the period of Industry 3.0, the implementation of the use of technologies that allowed for controlling everything from a physical terminal but programmed for that specific use that was created was evident. For example, the execution of automatic irrigation systems for crops, a robotic arm designed to assemble cars, or a computer that sends notifications when a product is about to run out of stock in the store would be examples of the search for automation (Mathur et al., 2022) or what can also be interpreted as the search for being able to implement the time of a human in some other task of the process.

A noticeable difference between the periods of the first two Industrial Revolutions is that they sought mechanization. At the same time, in Industry 3.0, the automation flag over the process is taken (Mathur et al., 2022). From a non-industrial perspective, it is understood that the efforts of various fields of knowledge generate epistemic content based on the progress they have on the knowledge they seek to generate (Cabrera-Ramírez & Cepeda Retana, 2022; Van-Gigch, 2022). For example, computer science has the purpose of independent application to Industry, but this does not mean it does not have applications for the same Industry. Being the evolution from mechanization to automation, a process developed by computer science, information technology, and related fields regarding technological development, it is possible to understand more clearly why organizations, when implementing knowledge about their processes that had this underlying goal, adopted these perspectives in their way of working and organizing (Elhusseiny & Crispim, 2022).

By automating, readings and data on the process are naturally obtained. By beginning to implement a series of technologies that can collect data that ultimately become helpful information for the process in which they have been applied, the same economic and industrial dynamics began to operate based on the analysis of the data obtained. Adding to the growing phenomenon of social networks in the last approximately 15 years of the 21st century, together with the constant connectivity that we achieved thanks to the implementation of the Internet in most devices, a "hyperconnectivity" era for humanity began gradually, and everything was connected thanks to the Internet (Vermesan, et al., 2022; Van Der Wees et al., 2022).

Here, the trend of the so-called "Internet of Things" (IoT) began to emerge. As previously mentioned, "things" not only read what they were applied to, but they also began to be connected to various communication networks among themselves (Van Der Wees et al., 2022), making things that were previously more complex, such as a real-time call between different parts of the world or a statistic of the most listened-to song in a particular region of the world, very accessible. According to the trend left by Industry 3.0 on the analysis of such data (hence the market trends for requiring trained personnel in data reading or "Big Data" knowledge), we now have total connectivity and digitalization of information in this environment that is present in most devices: the Internet (Alabadi et al., 2022).

This development was the next step in the idea of automation; however, the COVID-19 health pandemic accelerated the applications in the Industry given the impediment of being able to interact with large amounts of people or directly, the search to continue producing without the need for an in-person presence in the organizational processes by those involved (Kollu et al., 2022).

At this point, we observe the transition from the mere concept of "automating" to "digitizing" in Industry 4.0. In this period and its rapid transition due to the social and health context, the development of cyber-physical systems began, with operations in a "tangible" plane of reality but with information, procedures, and even products that do not escape the digital plane. We find implementing "clouds" of information along with IoT increasingly necessary to make decisions about processes, production, and industry development.

Industry 4.0 (I4.0) has been defined as the concept relating to the idea of the "industrial revolution", whose main objective is the integration of production processes with information technologies and techniques, combining production methods with information and communications technology (Nowotarski & Paslawski, 2017), in other words, the ongoing technological advancement of manufacturing components towards intelligent and interconnected systems (Niesen et al., 2016). This concept is also considered an umbrella term for a new industrial paradigm that embraces a set of future industrial developments regarding Cyber-Physical Systems (CPS), Internet of Things (IoT), Internet of Services (IoS), Robotics, Big Data, Cloud Manufacturing, and Augmented Reality (Pereira et al., 2017; Nagy et al., 2018), Artificial Intelligence, RFID, Embedded Systems, Virtual Reality and Machine Learning (Wichmann et al., 2019) to name a few. All technologies are intended to enable interoperability in the Industry, and it is unlikely that they will be deployed as individual technologies.

When the term Industry 4.0 is addressed, Cyber-Physical Systems are frequently found. This term refers to the interaction between the physical and the virtual environment, integrating, controlling, and coordinating processes and operations and, simultaneously, providing and using data accessing and processing (Pereira et al., 2017). CPS are systems in which different computational entities connect to the physical world and its processes collaborate. They provide and use, at the same time, access services and data processing available on the Internet (Monostori et al., 2016). In other words, CPS can be defined as "physical and engineered systems whose operations are monitored, controlled, coordinated, and integrated by a computing and communicating core". This kind of system CPS is a confluence of embedded systems, real-time systems, distributed sensor systems, and controls that will transform how humans interact with and control the physical world (Rajkumar et al., 2010).

It is important to emphasize that the production industry must have characteristics that allow the recovery, analysis, and processing of the data generated during production. On this, (Nowotarski & Paslawski, 2017) proposes five main characteristics of Industry 4.0: Interoperability, Virtualization, Decentralization, Real-Time Capability, and Modularity, which are explained below.

- **Interoperability:** Allows humans and smart factories[1] to connect and communicate.
- **Virtualization:** A virtual copy of the Smart Factory is created by linking sensor data with virtual plant and simulation models.
- **Decentralization:** Cyber-physical systems' ability to make their own decisions and produce locally, thanks to technologies such as 3d printing.
- **Real-Time Capability:** The capacity to collect and analyze data and provide the derived insights immediately.
- **Modularity:** Flexible adaptation of smart factories to changing requirements by replacing or expanding individual modules.

From all the above, we conclude that industrial revolutions can be understood in two large blocks: those that sought to mechanize the production force through the implementation of organizational processes entirely operated by people (1.0 and 2.0) and those that seek automation and digitization of processes through their connectivity, to facilitate the task that is performed within the process itself (3.0 and 4.0). Finally, some research indicates the advent of an "Industry 5.0" based on the use of Artificial Intelligence. However, the debate continues on whether this is another stage of Industry 4.0 being experienced in organizations or if it can really be talked about as a new way of focusing on production processes.

Internet of Things

The Internet of Things (IoT) concept has evolved since it was conceived in the late 1990s as part of Joy's six Webs taxonomy, where it was presented as a "device-to-device" (D2D) Web. This D2D Web refers to communication devices sharing information to manage, control and monitor different processes (Ibarra-Esquer et al., 2017). The term Internet of things also refers to the network interconnection of objects of various kinds, such as electronic devices and sensors, physical objects and human beings, data, and virtual environments. It arises from the trend of providing Internet connection to objects of daily life or everyday objects (Kosmatos et al., 2011).

Talking about IoT also requires talking about Ubiquitous Computing. This computer science area arises because of the increased processing capacity, storage capacity, miniaturization of computing devices, diversity and complexity of input and output devices, and the data transmission capabilities of networks. Furthermore, the incorporation of information technology beyond the confines of using tools towards a generalized penetration in daily life is studied (Araya, 1995). In 1991 Mark Weiser addressed integrating computers into everyday life until they are indistinguishable from it (Weiser, 1991); that is, people stop being aware that computers are present in their environment as they get used to them it. In other words, when incorporating hundreds of computers in a room, people use them unconsciously to people will use them unconsciously to accomplish everyday tasks. IoT and Ubicomp emerged in the 90s, and we have seen how they went from being ideas to reality today.

IoT devices can sense the environment and continuously collect data by transmitting it over a network. For this, IoT devices must have different general components that allow them to collect and transmit data, and seven characteristics of IoT devices stand out (Ibarra-Esquer et al., 2017):

- Have a physical embodiment and a set of associated physical features.
- Possess a unique identifier.
- Are associated with at least one name and one address.
- Can sense and store measurements made by sensor transducers associated with them.
- Have a minimal set of communication functionalities that allow them to make their identification, sensor measurements, and other attributes available to external entities such as other intelligent objects or systems.
- May possess means to trigger actions influencing the physical reality.
- Possess basic computing capabilities to make decisions about themselves and their interactions with external entities.

For the development of these characteristics, a systematic review of the changes that the IoT has undertaken since its conception and research on how technological advances have shaped it and fostered the arising of derived names suitable to specific domains enunciates five properties of IoT devices: Identification, Location and Tracking, Sensing, Actuation, and Processing (Ibarra-Esquer et al., 2017).

Identification refers to the fact that all IoT devices must have an associated name, identification number, or address to identify an intelligent object and establish an access path to the object. Location and Tracking mean that all individual objects will be tracked, and their condition and location will be communicated in real-time to a higher-level service. This concept not only refers to being aware of its physical location but also to its unique history from manufacturing to the end of its lifetime. Sensing property refers to the ability of things to collect data from the environment. Actuation refers to the influence of "Things" in their environment based on sensed data and controlled remotely via the Internet. Finally, the property of processing data and executing commands is frequently mentioned as intelligence or intelligent things with local intelligence and autonomy.

To develop Internet of Things applications, the two most common and widely considered architectures across the business, industrial research, and applications are three-layer (Zhang & Yu, 2013) and five-layer architectures, see Figure 2. The three-layer architecture consists of three layers: Perception layer, Network layer, and Application layer, while the five-layer architecture has the same three layers with two additional layers named access getaway layer and middleware layer (Al-Qaseemi et al., SA, 2017; Soumyalatha & Hegde, 2019).

Figure 2. Comparison between three-layer and five-layer architectures

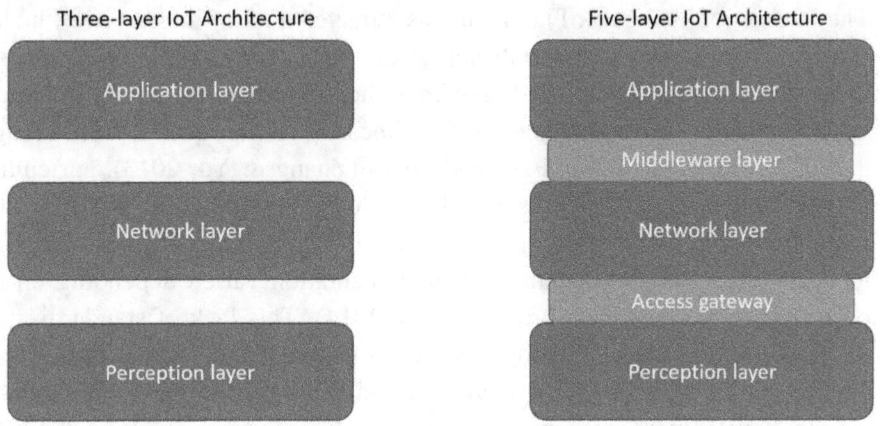

The perception layer is physical because it contains the physical devices and sensing layer due to embedded sensors in physical things. In this layer, embedded sensors collect the data from devices and send this data to the network layer. The network layer connects the perception layer to the physical layer. This layer must have connection technologies such as wired or wireless secure connection. Received data by the application layer will be analyzed and processed to provide services and make decisions, sending the results back from the application layer to the perception layer through the network layer. The access getaway layer manages IoT communication in the environment and exchanges messages between objects and systems. The middleware layer provides a more flexible association between hardware devices and applications (Al-Qaseemi et al., S. A, 2017). Despite the difference in the number of layers of both proposals, it is important to incorporate communication mechanisms between them to transmit the information adequately and achieve the required functionality.

IoT Devices

In recent years, commercializing everyday objects connected to the Internet has increased. According to the Statista website, in 2021, there was a total of 11.28 billion connected IoT devices, which is expected to double by 2027 (see Figure 3). This increase reflects the trend to incorporate this type of device more and more daily. These devices include home voice controllers like Google Home® or Amazon Echo®, smart TVs, smart lighting systems and light bulbs, smart plugs, thermostats, security cameras, door locks, and smartwatches. Nevertheless, IoT devices are currently being introduced in other areas, such as vehicles, through devices connected to the Bluetooth of the cell phone and the cell phone connected to the Internet.

The Internet of Things (IoT) domain has attracted a lot of interest over the last few years, to a large extent due to its applicability across a plethora of application domains (Ibarra-Esquer et al., 2017) like Smart homes (Khan et al. 2018) and Smart Cities (Righetti et al., 2018), medicine and healthcare (Farahani et al., 2018; Sadoughi et al., 2020), fire detection and fire industries (Zhang & Yu, 2013), agriculture (Gómez-Chabla et al., 2019; Kassim, 2020), education (Ane et al., 2020) and the Industry (Xu et al., 2014; Okano, 2017; Mondal, 2019; Wójcicki et al., 2022). It is noticeable that the architecture and implementation varied depending on the specific requirements of each project (Krco, 2014). This lack of standardization was a problem identified and addressed because it negatively affects the growth of the IoT and the security of IoT devices, causing an effort to have standards for the design, development, and implementation of the IoT. An example of this is the

Figure 3. Statistics of IoT devices
Source: www.statista.com

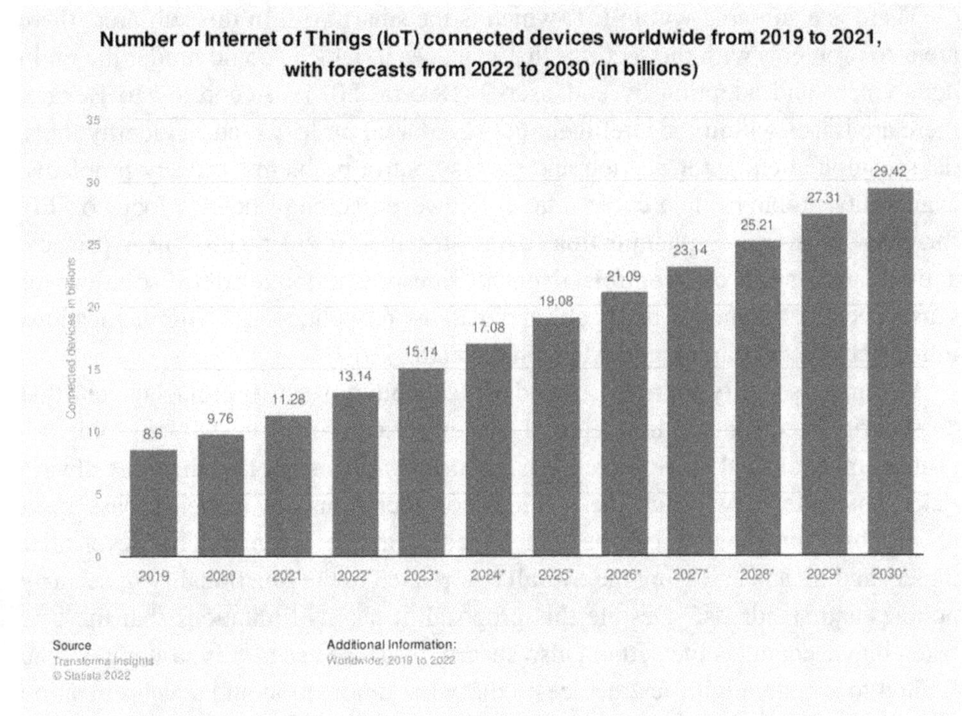

efforts made by the IEEE through the "IEEE Standards Activities in the Internet of Things (IoT) Overview" (IEEE Standards Association, 2023).

Security and Energy Consumption: Open Problems in IoT

Having devices connected to the network makes them vulnerable to possible attacks. IoT connections use existing network protocols since it "aims to connect objects on a large-scale using IP-based solutions (IP, TCP/UDP, etc.)" (Chakib Bekara, 2014), and IoT networks have experienced the development and standardization of a wide range of protocols to enable a wide range of IoT applications. This application includes wireless communication technologies, lightweight network management protocols, communication protocols for resource-constrained devices routing protocols for resource-constrained devices (Aboubakar et al., 2022). Implementing security processes in IoT devices is highly complex due to the limited capabilities available on the devices. Their vulnerabilities are related to transport encryption, inadequate software protection, and insufficient authorization. These risks threaten data privacy, which is crucial, especially when considering applications like health equipment and emergency services (Pereira et al., 2022).

There is a sub-area within IoT, which is the smart grid. In this sub-area, there are also problems with the security that manages it, which "could hinder the rapid deployment and adoption by end users" (Bekara, 2014). According to Bekara, there are issues within the intelligent network based on IoT, such as identity theft, data manipulation, authorization and access control problems, privacy problems, availability problems, and cyber-attacks. However, some principles focus on IoT for smart networks: authentication, data integrity, confidentiality, user privacy, authorization, and access control. Given the widespread adoption of IoT solutions for sensor networks, domotics, and health applications, providing secure communications in IoT networks is paramount (Meneghello et al., 2019).

Within the security approach around IoT technology, security proposals state that "if security is considered in the initial stage, they can solve substantial problems" (Tabassum et al., 2019). However, it is important to consider that there are always risks, no matter how much these initial considerations are made. Tabassum et al. propose that security can be "implemented at the organizational level after understanding and assessing the overall risk related to the functionality of security policies and standards." Despite this proposal, it has its limitations that must be taken into account, as the authors also specify that "the use of this strategy will be limited to a system with few devices; otherwise, automation and synchronization are required in the event of many devices that can help the system to avoid human errors," as it is proposed, there are others in which possible solutions are offered, but it must not be forgotten that security is a challenging point to carry out at your totality in any approach, since there are always possible risks of suffering severe consequences. Finally, security must be addressed throughout the device lifecycle, from the initial design to the operational environment (Lennvall,2017).

Energy

Considered by many as one of the most significant technological challenges for the following years; how energy is harvested, stored, and used will dictate the pace of evolution in the sector. Energy-efficient devices are imperative not only to enable more and more applications but also to make the existing ones greener and more environmentally friendly (Pereira et al., 2022). Besides, most of the existing IoT devices are low power, meaning they consume very little energy; on the other hand, some devices consume more due to the power they are handled, such as air conditioning and intelligent lighting. Within the energy consumption, an energy management system is proposed, which carries out the process that "collects energy measurement data from the field and makes it available in the form of visualization/graphs for users to easily understand when monitoring, analyze and manage energy" (Purwania et al., 2020), with which it is proposed that with this information with which an analysis can be carried out and used to generate strategy and decision systems for can lead to energy optimization.

Role of IoT in Industry 4.0

The Industrial Internet of Things (IIoT) should be understood as a subset of the broader IoT, where these connections exist mainly to produce physical goods for the marketplace and maintain the physical assets related to production (Mondal, 2019). This concept, according to Lampropoulos, is quite different from the general class of IoT, which accommodates a large number of nodes for real-time processing in Industry and combines several contemporary vital technologies to produce a system that functions more efficiently than the sum of its parts and focuses on automation, services, cloud computing, BD, CPSs and people; hence, IIoT can be considered as a complex system of diverse systems. (Lampropoulos et al., 2018) Furthermore, it blazes the trail to understanding the manufacturing process better, thus enabling efficient and sustainable production (Wójcicki,2022).

The IIoT uses IoT technologies in processes, products, and services that communicate with each other and with people over a global network. IoT-based devices, sensors, intelligent machinery, and equipment communicate through industry LAN or wireless communication systems; they can make manufacturing processes intelligent through self-optimization and autonomous decision applying Big Data platforms and Cloud Computing (Mondal, 2019). It has better-controlled latency at various hierarchical levels, provides more intelligent services in monitoring and alarm management with high reliability, high availability, and safe & resilience to failure, and it will create a new wave of disruptive companies which can be able to make automated decisions and take actions in real-time.

Challenges of IoT Adoption in the Industry

The Internet of Things in the Industry has the potential to contribute to the automation of regularly complex processes, as well as help to produce faster and reduce possible human errors through previously programmed devices and sensors. In addition, it is possible to activate equipment or machines and identify and detect the presence of people or vehicles near them automatically and remotely. It is essential to consider that "Most IoT solutions are designed for the consumer domain, so they do not support the harsh industrial requirements at all" (Lennvall et al., 2017). Applications have been devised in the Industry with devices with characteristics and capacities different from those in the home. This type of device in the Industry must offer availability, reliable and deterministic data communication, low latency, low jitter, and device interoperability with a high level of security" (Lennvall et al., 2017).

The automation pyramid is a general architecture proposed to assist in the automation of industries. This pyramid comprises sections or layers, and each specifies the processes that are selected based on their demand and importance and what devices can be included in each one. At the bottom is the field network, which comprises sensors and actuators; The next layer is the control network, which includes controllers and connectivity sensors; the next layer considers a client-server network; and the top layer considers a corporate network. At this last level are business decision systems. This automation pyramid can be observed in Figure 4.

Figure 4. Automation pyramid
Source: Adaptation of Lennvall et al. (2017)

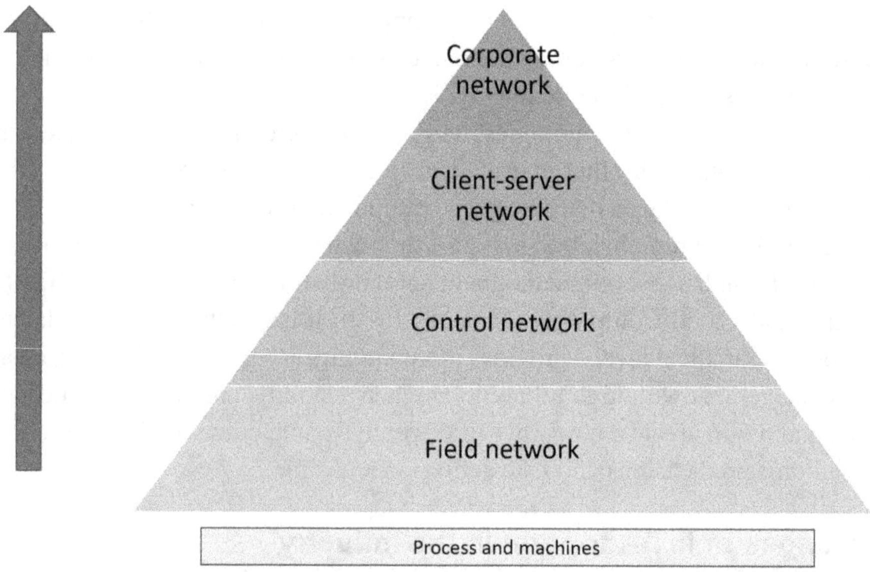

An extreme challenge is an adaptation that the IoT technology will have due to the technology that the industries already have, which is usually very old, and has already been inherited with the main designs that were previously available. It can also supplant and have better management in the processes carried out by the Industry. It is considered that the functions handled by the Industry are controlled by automation systems developed in each sector. For this reason, "the requirements of availability, determinism, and timely functional response are essential" (Lennvall et al., 2017).

For this reason, when implementing IoT technologies within the Industry, it must be considered that they must "provide a better or similar performance to that of current automation systems" (Lennvall et al., 2017). The integration between IoT systems, automation, and the feeding of information to a Programmable Logic Control (PLC) can cause problems that are related to security since quality is not associated with the automation system.

The hardware must be closely interconnected with Internet connectivity solutions. However, other connection models can be considered, such as Bluetooth ("consumer and industry will buy the IoT equipment and handle the connectivity themselves" (T. Lennvall, M. Gidlund and J. Åkerberg.2017)2017)), WiFi ("local connectivity is provided and managed by an IoT operator that charges the customer regularly" (T. Lennvall, M. Gidlund and J. Åkerberg. 2017)) and cellular operator ("global connectivity by 4G / 5G is provided by a mobile operator" (T. Lennvall, M. Gidlund and J. Åkerberg. 2017)).

Regarding its application in different industries, it is essential to recognize that methodologies may vary concerning the systems used to gather necessary information and automate existing processes to facilitate production. While the creation or definition of a methodology that can bring any organization from any industrial sector closer to the implementation of IIoT is beyond the scope of this writing, it is essential to establish general points that organizations should follow as application criteria to ensure proper assimilation of the process with the technology being implemented. To do this, general factors such as those presented in Figure 5 proposed by Ahmetoglu et al. (2022) must first be considered.

As seen in the previous chart, there are four fundamental pillars to consider during the implementation of IoT in any organizational process:

- **Taxonomic:** Refers to classification studies to detect the process's needs and what would help improve it.
- **Implementation Benefits:** This sets the goals and needs to be checked whether goals align with benefits to ensure optimal implementation.
- **Critical Factors:** This considers costs, unforeseen variables, and any other factor that may arise from the implementation.
- **Challenges:** This provides greater clarity on what the organization implementing these technologies will face, including technological, organizational, and contextual challenges (Ahmetoglu et al., 2022).

Figure 5. Factors for IoT adoption in organizations
Source: Ahmetoglu et al. (2022)

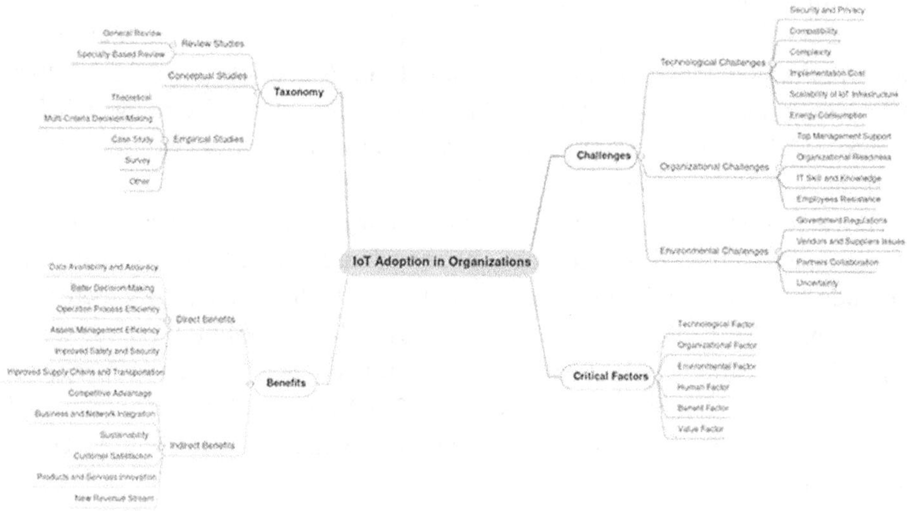

Subsequently, it is possible to analyze at a systemic level how this impacts the organizational structure in which IIoT is being applied, as shown in Figure 6, proposed by Anshari et al. (2022).

Figure 6. Factors for digital transformation
Source: Adaptation of Ansahri et al. (2022)

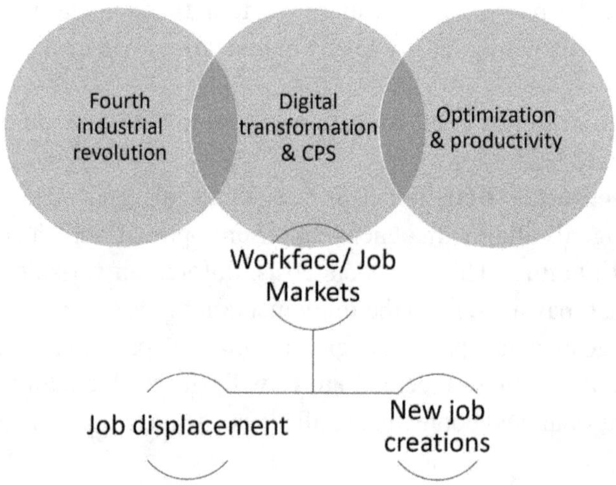

In this sense, IoT application in Industry is analyzed and implemented from a set of factors that function more as technical, theoretical, and application requirements than a transparent implementation methodology as continuous steps. This should not be surprising due to the vast number of applications in various industrial sectors to achieve constant connectivity by processes seeking to improve through the implementation of technologies. These general reviews are necessary to analyze all the impacts generated by this change in the methodology of production-oriented work (Anshari et al., 2022).

The IIoT would then take the following levels of requirements to operate correctly in a process according to Figure 7, proposed by Alabadi et al. (2022).

Figure 7. Dimensions of IoT application in an organization
Source: Alabadi et al. (2022)

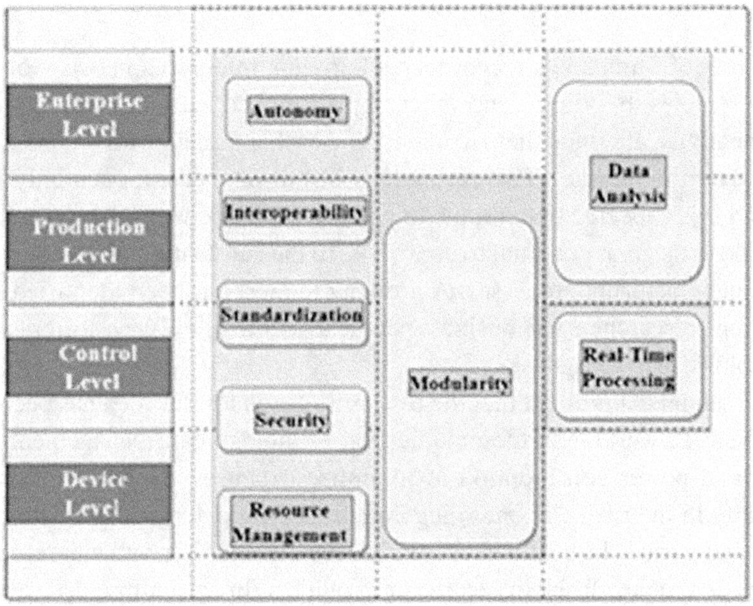

- **Enterprise:** The search for the autonomy of certain parts or the process as a whole.
- **Production Levels:** Refer to the interoperability between the process and the applied technologies.
- **Control Levels:** This encompasses process standardization due to its repeatability capacity and data security.
- **Devices:** This implies the management of the resource used for the technological and technical application of the process.

In these levels, data analysis and processing operate differently but systematically to address the process being followed. It should be remembered that productivity logic follows a series of steps and controls that establish the procedure to achieve the result.

Implementing IIoT requires a series of factors responding more to a situational analysis than a defined methodology. Examples of implementation can be found in different types of industries with particular methodologies in specific devices (Hang et al., 2022; Wójcicki et al., 2022); however, it is reiterated that the implementation of such technologies, at least for administrative and economic sciences, should be carried out according to situational analysis that responds to the factors defined in the information previously presented.

CONCLUSION

The Internet of Things is a recent technology that refers to a network of physical devices, vehicles, buildings, and other objects embedded with sensors, software, and connectivity, allowing them to collect and exchange data with other devices and systems over the Internet. This technology has developed exponentially in recent years, and it is expected to continue to grow in the same way in the coming years. This is due to its great potential to contribute to the automation of activities of daily life and manufacturing processes. A growing interest has been identified in recent years among researchers and business people to identify new ways to apply this type of technology to existing tasks.

It is essential to highlight that the research agenda for this area has been focused on improving networking and communication technology, data management, security, privacy, and power consumption to minimize expenses and increase feasibility, particularly in industries in emerging countries. The potential applications of IoT are vast and varied. In healthcare, IoT devices can monitor patients remotely and provide real-time feedback to healthcare providers. In agriculture, IoT sensors can optimize crop yields and conserve water. In manufacturing, IoT devices can monitor equipment performance and prevent downtime.

However, the proliferation of IoT devices also raises concerns about privacy, security, and data protection. As more devices connect to the Internet, the risk of cyber-attacks and data breaches increases. Manufacturers are developing more secure, privacy-focused IoT devices and protocols to address these concerns. Also, It is mandatory to establish good practices for designing IoT devices (i.e., smart devices) focusing on two main design challenges: power and connectivity.

REFERENCES

Aboubakar, M., Kellil, M., & Roux, P. (2022). A review of IoT network management: Current status and perspectives. *Journal of King Saud University - Computer and Information Sciences, 34*(7), 4163–4176. doi:10.1016/j.jksuci.2021.03.006

Ahmetoglu, S., Che Cob, Z., & Ali, N. A. (2022). A systematic review of Internet of Things adoption in organizations: Taxonomy, benefits, challenges and critical factors. *Applied Sciences (Basel, Switzerland), 12*(9), 4117. doi:10.3390/app12094117

Al-Qaseemi, S. A., Almulhim, H. A., Almulhim, M. F., & Chaudhry, S. R. (2016, December). IoT architecture challenges and issues: Lack of standardization. In *2016 Future technologies conference (FTC)* (pp. 731–738). IEEE. doi:10.1109/FTC.2016.7821686

Alabadi, M., Habbal, A., & Wei, X. (2022). Industrial internet of things: Requirements, architecture, challenges, and future research directions. *IEEE Access : Practical Innovations, Open Solutions, 10*, 66374–66400. doi:10.1109/ACCESS.2022.3185049

Ane, T., Billah, M., & Nepa, T. (2020). Performance of Internet of things (IoT) potential applications in education. *Bangladesh Journal of Multidisciplinary Scientific Research*, *2*(2), 10–16. doi:10.46281/bjmsr.v2i2.653

Anshari, M., Syafrudin, M., & Fitriyani, N. L. (2022). Fourth Industrial Revolution between Knowledge Management and Digital Humanities. *Information (Basel)*, *13*(6), 292. doi:10.3390/info13060292

Araya, A. A. (1995, February). Questioning ubiquitous computing. In *Proceedings of the 1995 ACM 23rd annual conference on Computer science* (pp. 230–237). 10.1145/259526.259560

Ashton, K. (2009). That 'Internet of things' thing. *RFID Journal*, *22*(7), 97–114.

Barrientos-Avendaño, E., Areniz-Arevalo, Y., Coronel-Rojas, L. A., Cuesta-Quintero, F., & Rico-Bautista, D. (2020). Modelo de incursión en la industria 4.0 aplicado a la compañía alimenticia tu pan gourmet SAS: Estrategia para el renacer en la pandemia ocasionada por COVID-19 (SARS-CoV-2). *Revista Ibérica de Sistemas e Tecnologias de Informação*, (E34), 436–449.

Becerra, W. J. A., Luna, A. R. L., & Aponte, G. C. V. (2021). Efectos organizacionales durante la pandemia covid-19: El caso de conamovidi y los comedores autogestionarios. *Revista de Sociologia (Lima)*, (33), 61–76. doi:10.15381/rsoc.n33.21795

Bekara, C. (2014). Security issues and challenges for the IoT-based smart grid. *Procedia Computer Science*, *34*, 532–537. doi:10.1016/j.procs.2014.07.064

Blanc, R., Ratto, D., Cettour, W., & Lepratte, L. (2021). *Modelos de madurez e implementación en industria 4.0 análisis de alternativas y nivel de implementación*. Academic Press.

Cabrera-Ramírez, S., & Cepeda-Retana, J. (2022). La epistemología, guía para el conocimiento científico. *Portal de la Ciencia, 3*(2), 123–133. doi:10.51247/pdlc.v3i2.317

Cano-Orón, L., & Llorca, G. (2018). La configuración invisible del espacio digital: el caso de Facebook. *Observatorio (OBS*), 12*(3), 106-121.

Choi, T. M., Kumar, S., Yue, X., & Chan, H. L. (2022). Disruptive technologies and operations management in the Industry 4.0 era and beyond. *Production and Operations Management, 31*(1), 9–31. doi:10.1111/poms.13622

Clavijo-Quintero, D. P. (2022). *Análisis de la Industria 4.0 En América Latina: Caso México y Colombia*. Academic Press.

Coyoli-Pereyra, D. M. (2022). *La calidad de vida a través de los sentimientos en las revoluciones industriales*.

Da Xu, L., He, W., & Li, S. (2014). Internet of things in industries: A survey. *IEEE Transactions on Industrial Informatics, 10*(4), 2233–2243. doi:10.1109/TII.2014.2300753

Elhusseiny, H. M., & Crispim, J. (2022). SMEs, Barriers and Opportunities on adopting Industry 4.0: A Review. *Procedia Computer Science, 196*, 864–871. doi:10.1016/j.procs.2021.12.086

Farahani, B., Firouzi, F., Chang, V., Badaroglu, M., & Constant, N. (n.d.). *Towards fog-driven IoT eHealth: Promises and challenges of IoT in medicine and healthcare*. Future20. doi:10.1016/j.future.2017.04.036

Ferrer, A. (2022). *Historia de la globalización II: La Revolución Industrial y el segundo orden mundial*. Fondo de Cultura Económica Argentina.

García, R. C., & Oliva, M. A. (2021). La transformación digital de los medios de pago: Impactos de la covid-19. In Managing the future: Challenges and Proposals for Post-pandemic Society (pp. 243-253). Universitat Rovira i Virgili.

Gómez, R. R. (2022). Internet de las cosas, salud pública y pandemias. *Salud (Barranquilla), 38*(1), 240–253. doi:10.14482un.38.1.614.4

Gómez-Chabla, R., Real-Avilés, K., Morán, C., Grijalva, P., & Recalde, T. (2018, December). IoT applications in agriculture: A systematic literature review. In *ICT for Agriculture and Environment: Second International Conference, CITAMA 2019, Guayaquil, Ecuador, January 22-25, 2019, Proceedings* (pp. 68-76). Cham: Springer International Publishing. 10.1007/978-3-030-10728-4_8

Guerrero, D. S. (2022). El debate en torno a la noción instrumental de tecnología. *Revista Ethika*, (6), 149–168. doi:10.5354/2452-6037.2022.66375

Hang, N. T., Huy, D. T. N., Le, T. H., Gwoździewicz, S., Thanh, N. T. P., Dung, N. T., & Hien, D. T. N. (2022). Further Analysis on Internet of Things (IOT) Applications in Emerging Markets and Vietnam. In Ambient Communications and Computer Systems. *Proceedings of RACCCS, 2021*, 407–416.

Ibarra-Esquer, J. E., González-Navarro, F. F., Flores-Rios, B. L., Burtseva, L., & Astorga-Vargas, M. A. (2017). Tracking the evolution of the Internet of things concept across different application domains. *Sensors (Basel)*, *17*(6), 1379. doi:10.339017061379 PMID:28613238

IEEE Standards Association. (2023, February 14). *Internet of things*. IEEE Standards Association. Retrieved March 23, 2023, from https://standards.ieee.org/initiatives/iot/

Iqbal, M. J., & Khizar, H. M. U. (2022). The epistemology of entrepreneurial responsible orientation (ERO): Theory, conceptualization, and future research. *SAGE Open*, *12*(2). doi:10.1177/21582440221089320

Jiang, Z., Yuan, S., Ma, J., & Wang, Q. (2022). The evolution of production scheduling from Industry 3.0 through Industry 4.0. *International Journal of Production Research*, *60*(11), 3534–3554. doi:10.1080/00207543.2021.1925772

Kassim, M. R. M. (2020, November). IoT applications in smart agriculture: Issues and challenges. In *2020 IEEE conference on open systems (ICOS)* (pp. 19-24). IEEE. doi:10.1109/ICOS50156.2020.9293672

Khan, A., Al-Zahrani, A., Al-Harbi, S., Al-Nashri, S., & Khan, I. A. (2018, February). Design of an IoT smart home system. In *2018 15th Learning and Technology Conference (L&T)* (pp. 1–5). IEEE. 10.1109/LT.2018.8368484

Kollu, P. K., Kumar, K., Kshirsagar, P. R., Islam, S., Naveed, Q. N., Hussain, M. R., & Sundramurthy, V. P. (2022). Development of advanced artificial intelligence and IoT automation in the crisis of COVID-19 Detection. *Journal of Healthcare Engineering*.

Kosmatos, E. A., Tselikas, N. D., & Boucouvalas, A. C. (2011). Integrating RFIDs and Smart Objects into a UnifiedInternet of Things Architecture. *Advances in Internet of Things*, *01*(01), 5–12. doi:10.4236/ait.2011.11002

Krčo, S., Pokrić, B., & Carrez, F. (2014, March). Designing IoT architecture (s): A European perspective. In *2014 IEEE World Forum on Internet of Things (WF-IoT)* (pp. 79-84). IEEE. doi:10.1109/WF-IoT.2014.6803124

Laffi, M., & Boschma, R. (2022). Does a local knowledge base in Industry 3.0 foster diversification in Industry 4.0 technologies? Evidence from European regions. *Papers in Regional Science, 101*(1), 5–35. doi:10.1111/pirs.12643

Lampropoulos, G., Siakas, K., & Anastasiadis, T. (2018). Internet of Things (IoT) in Industry: Contemporary Application Domains, Innovative Technologies and Intelligent Manufacturing. *International Journal of Advances in Scientific Research and Engineering, 4*(10), 109–118. doi:10.31695/IJASRE.2018.32910

Lennvall, T., Gidlund, M., & Akerberg, J. (2017). Challenges when bringing IoT into industrial automation. 2017 IEEE AFRICON: Science. *Technology and Innovation for Africa, AFRICON, 2017*, 905–910. doi:10.1109/AFRCON.2017.8095602

López-Pérez, E. (2020). *La industria 4.0 y las nuevas formas de trabajar: una perspectiva desde el caso mexicano en tiempos del COVID 19*. Academic Press.

Madakam, S., Lake, V., Lake, V., & Lake, V. (2015). Internet of Things (IoT): A literature review. *Journal of Computer and Communications, 3*(05), 164–173. doi:10.4236/jcc.2015.35021

Mathur, A., Dabas, A., & Sharma, N. (2022). Evolution From Industry 1.0 to Industry 5.0. In *2022 4th International Conference on Advances in Computing, Communication Control and Networking (ICAC3N)* (pp. 1390-1394). IEEE.

Medina-Valdés, Z., Díaz-Roque, J., & Plaza-Macías, N. M. (2022). La Cuarta Revolución Industrial y el sistema de contradicciones del capitalismo. *Economia y Desarrollo, 166*(1).

Meneghello, F., Calore, M., Zucchetto, D., Polese, M., & Zanella, A. (2019). IoT: Internet of Threats? A Survey of Practical Security Vulnerabilities in Real IoT Devices. *IEEE Internet of Things Journal, 6*(5), 8182–8201. doi:10.1109/JIOT.2019.2935189

Mondal, D. (2019). The Internet of Things (IoT) and Industrial Automation: A future perspective. *World Journal of Modelling and Simulation, 15*(2), 140–149. Retrieved from www.rcciit.org doi:10.1016/j.cirp.2016.06.005

Monostori, L., Kádár, B., Bauernhansl, T., Kondoh, S., Kumara, S., Reinhart, G., ... Ueda, K. (2016). Cyber-physical systems in manufacturing. *CIRP Annals, 65*(2), 621-641. doi:10.1016/j.cirp.2016.06.005

Muñoz, J, C. (2022). *Sobre la enseñabilidad de la administración. Algunas reflexiones preliminares*. Academic Press.

Nagy, J., Oláh, J., Erdei, E., Máté, D., & Popp, J. (2018). The role and impact of industry 4.0 and the internet of things on the business strategy of the value chain-the case of Hungary. *Sustainability (Basel), 10*(10), 3491. Advance online publication. doi:10.3390u10103491

Niesen, T., Houy, C., Fettke, P., & Loos, P. (2016, January). Towards an integrative big data analysis framework for data-driven risk management in industry 4.0. In *2016 49th Hawaii international conference on system sciences (HICSS)* (pp. 5065–5074). IEEE. 10.1109/HICSS.2016.627

Nowotarski, P., & Paslawski, J. (2017, October). Industry 4.0 concept introduction into construction SMEs. In IOP Conference Series: Materials Science and Engineering (Vol. 245, No. 5, p. 052043). IOP Publishing. doi:10.1088/1757-899X/245/5/052043

Okano, M. T. (2017). IOT and Industry 4.0: The Industrial New Revolution. *ICMIS-17 - International Conference on Management and Information Systems*, 75–82.

Papadopoulos, T., Singh, S. P., Spanaki, K., Gunasekaran, A., & Dubey, R. (2022). Towards the next generation of manufacturing: Implications of big data and digitalization in the context of industry 4.0. *Production Planning and Control, 33*(2-3), 101–104. doi:10.1080/09537287.2020.1810767

Park, S. (2016). Development of innovative strategies for the Korean manufacturing industry by use of the Connected Smart Factory (CSF). *Procedia Computer Science, 91*, 744–750. doi:10.1016/j.procs.2016.07.067

Pereira, A. C., & Romero, F. (2017). A review of the meanings and the implications of the Industry 4.0 concept. *Procedia Manufacturing, 13*, 1206–1214. doi:10.1016/j.promfg.2017.09.032

Pereira, F., Correia, R., Pinho, P., Lopes, S. I., & Carvalho, N. B. (2020). Challenges in resource-constrained IoT devices: Energy and communication as critical success factors for future IoT deployment. *Sensors (Basel), 20*(22), 6420. doi:10.339020226420 PMID:33182749

Purwania, I. B. G., Kumara, I. N. S., & Sudarma, M. (2020). Application of IoT-Based System for Monitoring Energy Consumption. *International Journal of Engineering and Emerging Technology, 5*(2), 81–93.

Rajkumar, R. R., Lee, I., Sha, L., & Stankovic, J. (2010). Proceedings of the 47th design automation conference. *Cyber-physical systems: The next computing revolution.* 10.1145/1837274.1837461

Reynaldos-Grandón, K. (2020). Cambio y Liderazgo: Aprendizaje organizacional en tiempos de pandemia. *Revista Medica de Chile, 148*(9), 1375–1377. doi:10.4067/S0034-98872020000901375 PMID:33399719

Righetti, F., Vallati, C., & Anastasi, G. (2018, June). IoT applications in smart cities: A perspective into social and ethical issues. In *2018 IEEE International Conference on Smart Computing (SMARTCOMP)* (pp. 387-392). IEEE. 10.1109/SMARTCOMP.2018.00034

Sadoughi, F., Behmanesh, A., & Sayfouri, N. (2020). Internet of things in medicine: a systematic mapping study. *Journal of Biomedical Informatics*, 103. doi:10.1016/j.jbi.2020.103383

Sánchez-Moncayo, G. K. (2021). *El impacto económico y el uso de la transición digital como innovación en el sector minorista de comercios especializados y no especializados frente a la pandemia COVID–19 en la zona centro del país* [Bachelor's thesis]. Universidad Técnica de Ambato. Facultad de Ciencias Administrativas. Carrera de Organización de Empresas.

Soumyalatha, S. G. H. (2016, May). Study of IoT: understanding IoT architecture, applications, issues and challenges. *International Journal of Advanced Networking & Applications, 478.*

Tabassum, K., Ibrahim, A., & El Rahman, S. A. (2019, April). Security issues and challenges in IoT. In *2019 International Conference on Computer and Information Sciences (ICCIS)* (pp. 1–5). IEEE.

Van Der Wees, A., Breeuwsma, J., & van Sleen, A. (2022). IoT Societal Impact–Legal Considerations and Perspectives. In Digitising the Industry Internet of Things Connecting the Physical, Digital and VirtualWorlds (pp. 215-235). River Publishers. doi:10.1201/9781003337966-7

Van-Gigch, J. P. (2002). Comparing the epistemologies of scientific disciplines in two distinct domains: modern physics versus social sciences. II: epistemology and knowledge characteristics of the 'new'social sciences. *Systems Research and Behavioral Science, 19*(6), 551–562. doi:10.1002res.466

Vermesan, O., EisenHauer, M., Serrano, M., Guillemin, P., Sundmaeker, H., Tragos, E. Z., ... Darmois, E. C. (2022). The next generation internet of things–hyperconnectivity and embedded intelligence at the edge. In Next Generation Internet of Things–Distributed Intelligence at the Edge and Human-Machine Interactions (pp. 19-102). River Publishers.

Weiser, M. (1991). The Computer for the 21st Century. *Scientific American, 265*(3), 94–105. doi:10.1038cientificamerican0991-94 PMID:1675486

Wichmann, R., Eisenbart, B., & Gericke, K. (2019). The Direction of Industry: A Literature Review on Industry 4.0. *Proceedings of the Design Society: International Conference on Engineering Design, 1*(1), 2129–2138. doi:10.1017/dsi.2019.219

Wójcicki, K., Biegańska, M., Paliwoda, B., & Górna, J. (2022). Internet of Things in Industry: Research Profiling, Application, Challenges and Opportunities—A Review. *Energies, 15*(5), 1806. Advance online publication. doi:10.3390/en15051806

Zhang, Y. C., & Yu, J. (2013). *A study on the fire IOT development strategy*. Procedia Engineering. doi:10.1016/j.proeng.2013.02.146

Zizic, M. C., Mladineo, M., Gjeldum, N., & Celent, L. (2022). From industry 4.0 towards industry 5.0: A review and analysis of paradigm shift for the people, organization and technology. *Energies, 15*(14), 5221. doi:10.3390/en15145221

KEY TERMS AND DEFINITIONS

Administrative and Economic Sciences: Academic disciplines that study the organization, planning, direction, and control of resources in companies and institutions, and the economic relationships between agents.

Automation: Application of technology to perform task and processes with minimal human intervention.

Industry: Economic sector that is dedicated to the production and commercialization of goods and services.

Industry 4.0: Use of technology in the industry aimed to automate processes and reduce operational costs.

Industrial Internet of Things (IIoT): Use of IoT technology in industrial settings to improve efficiency, reduce costs, and increase productivity.

Industrial Revolution: Economic and social transformation that occurred between the late 18th and early 19th centuries, in which machinery and mass production techniques were introduced.

Internet of Things (IoT): Technology that connects everyday objects to the internet, allowing them to interact with the environment and communicate with each other.

Organizational Process: Activities and tasks that an organization performs to achieve its objectives, such as planning, execution, and control.

Production: Process of creating goods and services, from planning to delivery to the end consumer.

ENDNOTE

1 A hyper-connected network-based integrated manufacturing system that acquires all information on manufacturing facilities in real time through the Internet, autonomously changes a manufacturing method, replaces raw materials, and ultimately implements an optimized dynamic production system (Park, 2016).

Chapter 7
Cyber Security:
Trends and Appraisal on Threats, Attacks, and Security Models

Priyatosh Jana
https://orcid.org/0000-0002-5219-2089
Haldia Institute of Technology, India

Debjit Banerjee
Haldia Institute of Technology, India

Krishnashis Das
Haldia Institute of Technology, India

Soham Maity
Haldia Institute of Technology, India

Abhijit Sarkar
https://orcid.org/0000-0002-8361-9057
Haldia Institute of Technology, India

Sabyasachi Samanta
https://orcid.org/0000-0001-8858-4132
Haldia Institute of Technology, India

ABSTRACT

Both cyber security and information security are frequently used interchangeably. The latter adds a new dimension to the former by considering human variables throughout the security process, making the individual the focus of attention and a potential target. However, such talks about cyber security often revolve around social facets of society and have significant implications. To aid in cyber security, several models and systems have been developed. Both the number and complexity of hacks have increased dramatically over the past few decades. Because of this, developing a plan that cannot be compromised by online assaults is essential. As hackers become more sophisticated in their use of state-of-the-art techniques and cutting-edge technology, we are seeing an increase in hacking, attacks, and data leaks. In this chapter, the authors investigate few models and recent trends of threats and attacks of security models. In addition, they have also studied their shortcomings and the methods previously employed to counteract such hazards.

DOI: 10.4018/978-1-6684-8639-9.ch007

INTRODUCTION

Cyberspace is the dynamic environment that arises from the interaction of machines, software, and services. The spread of laptops and other types of technological contact sustains globalisation. It is challenging to create clear boundaries between private people, businesses, military, and governmental organisations in online due to the benefits given by technological advancement. Cyberspace is only anticipated to become more complicated in the coming decades as more and more networks and devices become linked to one another.

The increasing prevalence of cyber dangers and -attacks has made cyber security an urgent concern. This is because cybercriminals have developed more sophisticated strategies for targeting computer systems. Individuals, start-ups, and larger businesses are all impacted. Organizations of all hues, not just those in the IT sector, are starting to wake up to the seriousness of the situation and take measures to strengthen their defences against cybercrime. Cyber security is primarily about individuals, procedures, and techniques working together to mitigate threats and vulnerabilities through deterrence, international involvement, incident response, resilience, and restoration using tools like computer network operational processes, information assurance, law enforcement, etc. The importance of cyber security in today's highly digital world cannot be overstated. Cyberattacks can be extremely costly for businesses. Damage to a company's image and bottom line can result from a data leak. The harm caused by cyberattacks in the modern era is mounting. Cybercriminals' elaborate online attacks evolve over time (Maglaras et al., 2022).

The two most common forms of hacking are denial-of-service and insider threats. Threats posed by the Internet Attacks on the system, subsequently. An attack on a website or its related online apps that originates from within the Internet itself. Some frequent instances of internet violence are as follows:

As a starting point, we have injection attacks, which happen when hackers insert harmful code or data into a website in an effort to gain access to sensitive data. SQL Injection, code injection, log injection, XML injection, and so on are just a few of the most prevalent types of injection attacks.

One vulnerable area of computer network security is DNS hijacking. DNS cache poisoning attacks aim to redirect network traffic to an unauthorised address by inserting harmful content into a DNS resolver's cache. DNS faking assaults pose a serious threat to security because they may go unnoticed for long periods of time.

Finally, in regards to protected networks, a session hacking assault is a security violation that specifically targets a user's session. Web applications use "cookies," which are text files, to remember information about a user's experience and their preferences. If a thief obtains access to your cookies, they can access any information stored in them, no matter how private.

Phishing is an aggressive method of obtaining personal information from users, such as passwords and credit card numbers. It occurs when a criminal masquerades as a safe person you can communicate with online.

Physical force is a try-and-see approach to offence. This method generates and tests a large number of possibilities in order to crack a passcode or ID number. While hackers may employ this attack to gain access to protected data, security experts may employ it to gauge the efficacy of a company's network defences.

The goal of the sixth type of attack is to prevent users from accessing a website or other network resource. To accomplish this, it floods the target with traffic or transmits it corrupt data that causes it to collapse. It uses a single machine and a single internet link to launch a coordinated assault against a website. Because of this, it can be classified as one of the following: A strategy of overwhelming force the goal of the attack is to consume as many data per second as possible from the targeted website. Attacks against a protocol can be measured in terms of the amount of packets sent and received and the actual amount of server resources consumed. Application layer attacks are those that aim to flood a web server with queries (Khan et al. 2022).

As another common technique, dictionary assaults store a list of commonly used passwords and use authentication to figure out the real one. In some cases, an intruder can trick a web server into displaying material that the user is not authorised to view by tampering with the Address. Putting meaning into an Address is what that term describes.

At number nine, we have file inclusion attacks, which use the include function to read or run any file on the web server. This includes both text files and important system files. Attack type ten is a "man in the middle" assault, in which the perpetrator acts as a "middleman" between the customer and the host. An attacker can then read, insert, and otherwise alter the data within the captured link (Veksler et al., 2021).

Because of regulations like the General Data Protection Regulation, businesses are strengthening their data protection procedures (GDPR). Because of these factors, cyber security is now an integral part of contemporary companies, and the focus has moved to developing efficient reaction strategies to mitigate any damage that may occur.

Malicious hacking attacks have become more complex as cybercriminals learn to exploit vulnerabilities in both machines and their human targets. While this is true, a number of precautions in the form of monitoring and avoidance processes have been created. This paper aims to do two things: (1) provide a comprehensive overview of study on cyber assault techniques, discovery, and avoidance; and (2) cast light on the many aspects of cyber security issues. Therefore, people need to be taught about cyber-attack, discovery, and prevention methods as a preventative step (Mathew, 2019).

Here we have performed a comprehensive discussion in recent trends of cyber treats, cyber-attacks and cyber security models. In addition we have explored few recent models in cyber security by introducing the conception of information security and network security. Professionals listed a number of reasons for cyber-crime in modern days. Lack of awareness, lack of technical knowledge among police and people and growth of online traffic are the very common intention for this. Practice of creating fake emails or SMS (known as 'Phishing') and target it to common users are very common practice as they trust you. Malware, Social Engineering, Viruses, Email Worm, Internet Worms, Logic Bombs, Spyware, Ransom Ware etc. are also another kind of digital cyber-crime in recent days. Recent trends on the above issues including change of cyber security market of last few years are also comprehensively depicted in this work also.

LITERATURE REVIEW

Li et al. (2021) proposed a broad analysis of cyber-attacks and cyber security. They also comprised some emerging trends, developments of cyber security and security threats with their challenges. Ravi Sharma has presented some recent cyber security issues. A recent cyber security attack possible in mobile devices, social media networking, cloud computing etc. was discussed. Lee et al. (2016) introduced an innovative authentication technique using mouse clicks. In this work, it was assessed how vulnerable mouse data is using readily available mouse loggers. The studied result may be used to create a security policy that will protect authentication data based on picture recognition from being leaked. Daniel Mellado et al. (2014) has suggested that a goal-driven security requirements engineering technique, like Secure Tropos, is a good paradigm for eliciting security needs and analysing them from both a social and a technological perspective. Yet, previous approaches to SPL engineering have routinely overlooked security requirements, and goal-driven security requirements engineering processes are not well suited to the unique challenges of SPL.

Using the "Secure Tropos-SPL" architecture, this is proposed in this study. IoT Checker, suggested by Mujahid Mohsin et al. (2017), is a cutting-edge data-driven framework for modelling IoT setups with the goal of automatically detecting security configuration irregularities and analysing IoT-specific attack vectors. The method makes use of both freshly built ontologies and updated, standardised versions of pre-existing ones. By using rules-supported reasoning and queries to characterise the context of complex IoT interactions and dependencies, configuration analytics may be carried out automatically. Assessment is achieved by ontology-based security categorization of 954 commercially available Internet of Things (IoT) devices and security analysis of their in-use system configurations. The fully automated method has been shown to be scalable, readily controllable, formally verifiable, and devoid of mistakes caused by laborious human setups.

Frank Cremer et al. (2022) proposed a review on Cyber risk and cyber security. The study is very helpful for to comprehend, metricize and manage cyber risks. Kangbin Yim Et al. (2010) provided practical solutions to the hardware vulnerability of the keyboard controller. This flaw makes it feasible for an attacker to steal passwords even if keyboard protection software is enabled. Even if an attacker is able to sniff the keyboard hardware communication, the suggested strategy will make it difficult for them to get the actual password. An asymmetric cryptosystem may be developed for secure communication based on Chebyshev polynomials. Composition of these polynomials forms a semi-group, which is the basis of this cryptosystem. Two novel cryptosystems are introduced by Marcin Lawnik et al. (2019), both of which use variants of Chebyshev's polynomials. The study provided here demonstrates that their level of safety is equivalent to that of methods related to the discrete logarithm issue. Chebyshev polynomials may be computed quickly using the techniques shown in the article.

Divesh Aggarwal et al. (2016) have presented that under the generic ring model of computing, where an algorithm may execute ring operations, inverse ring operations, and test equality, the challenge of factoring N can be efficiently simplified to solving the RSA problem on Z_N. This demonstrates that the RSA encryption and digital signature technique is secure against some cryptanalytic attacks, including those that rely on the factoring assumption.

In a recent article, Md. Liakat Ali et al. (2020) provided a high-level review of the cyber security issues brought about by technological progress and new ideas. Modern approaches, trends, and other cyber security ethics are also the topic of this article. A number of models and frameworks have been developed with the aim of improving cyber security.

Kutub Thakur et al. (2015) provides an introduction to cyber security, outlining the fundamentals of the field and discussing the people that work in this field as well as the resources necessary to ensure the safety of sensitive data stored digitally. In this study, we examine these models, their shortcomings, and the methods previously used to counteract these dangers. Further suggestions for additional studies are included in the paper.

Jing Li (2015) has investigated the issues of firewall placement in network topology during design and routing table construction during operation with the goal of reducing the maximum number of firewall rules. There are two key contributions made by the author. It begins by demonstrating that these problems are in fact NP-complete. The author then provide a heuristic method and use simulations to prove the technique works. The findings demonstrate that the maximum multi-firewall rule set is reduced using the suggested technique compared to previous algorithms.

CYBER SECURITY

Cyber security is the practise of protecting data, computers, servers, mobile devices, electronic systems, networks, and other digital assets against harm (such as fraud or unauthorised access). Given the prevalence of electronic storage and transmission of sensitive data, this is an essential component of contemporary technology.

Information security encompasses a variety of technologies, processes, and practices to ensure the confidentiality, integrity, and availability of information. Here are some typical examples:

Firewall: Hardware or software that acts as a barrier between a computer or network and the Internet, protecting against unauthorised access and malicious traffic.

Encryption: The process of converting information into an encrypted format to protect it from unauthorised access.

Antivirus Software: Software that detects and removes malicious software, such as viruses, from a computer or network.

Access Control: Measures that restrict access to a computer system or network, such as user authentication and authorization.

Network Security: Measures to protect the security of computer networks, such as Network segmentation and virtual private networks (VPNs).

Disaster Recovery Planning: Procedures for restoring systems and data in the event of a computer attack or natural disaster.

Security Awareness Training: Programs that educate employees and the public about cybercrime threats and best practices for protecting against them.

Cyber security is important as the increasing use of technology has led to an increase in cyber-attacks that can lead to theft of confidential information, financial loss and reputational damage. In addition, cyberattacks on critical infrastructure such as power grids or communication networks can cause widespread outages and even pose a threat to national security.

To ensure effective cybersecurity, it is important to keep up to date with the latest security technologies and practices and to continuously evaluate and improve existing security measures. This requires a combination of technical solutions and security best practices, as well as a commitment to ongoing education and awareness of the ever-changing threat landscape. (Ramirez, R.ET al., 2016). Here in Figure 1 the different types of cyber-attacks are presented.

Figure 1. Different types of cyber attacks

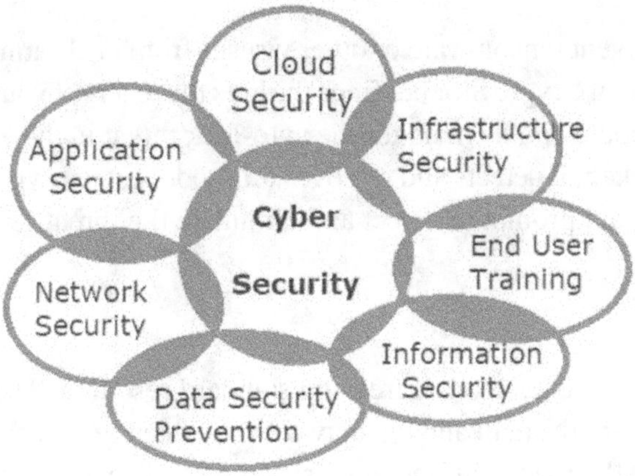

CYBER SECURITY TOOLS

Here, we take a look at the state-of-the-art cryptographic solutions to the myriad of significant cybercrimes that may be prevented, or at least minimized, by using cyber security measures. Some of the best cybersecurity tools available today are:

BMQ_Radar Advisor and Watson

As far as corporate security measures go, this is the gold standard. Watson is an AI that is constantly adapting and learning. Here's how it functions: Code like this is used by IBM Q-Radar to track down and compile data and connections from internet, offline, and local sources. Prepare a plan to conceal the truth, and then eliminate the danger when it arises. One of the finest things that may happen to you online is if you manage to disable your own protection measures.

Wire Shark

It's a standard method for analysing networks. Determine the level of risk associated with the user's current network location. There have been reports of Wireshark either logging or displaying information about network use. The protocol utilised in transmission, as well as incoming and outgoing data packets, are all plainly viewable. Collect data in real time and use an offline analysis sheet to look it over afterwards.

Crypto Stopper

If you want to prevent ransomware or other viruses from infiltrating your computer, this is a great option. Crypto Stopper finds the bots that encrypt your files and deletes them. Create a model and formula for the auto-lock threat in the formula. Once the crypto lock is locked, it detects and removes this code. An encryption block usually promises a 9-second prompt to detect and eliminate the threat.

N Map

For titles on the network, this is a free and standard resource. The effectiveness of NMAP extends to both small and big networks. Locate the sender and the recipient in the network. All versions of operating systems are supported, too.

Burp Suite

Here's another web application security tool that uses algorithms to search the web. The primary goals of this instrument are system control and penetration. All sender and target requests and replies to the threat should be examined. Finding a danger and eliminating it is possible.

Open VAS

Similar to the Nessus tool but extremely different from Nessus and Metasploit in appearance and operation. Today it is considered one of the most stable, least vulnerable and most widely used network security tools on the Internet.

There are two main components of Open VAS:

Scanner: Scans sensitive sections and sends a compiled report of everything to its manager.
Manager: Compile all requests received from the scanner and then report all these events.

CHANGING TRENDS IN CYBER SECURITY

Here are some of the trends that are having a huge impact on cyber security:

Web Servers

Still, attacks on online apps to collect data or propagate malware are a real concern. Hackers use compromised legitimate web servers to spread malware. However, data theft attacks, which are becoming common and hence drawing more public attention, also represent a serious risk. As a result, we must now place a greater priority on ensuring the safety of web servers and online applications. Cybercriminals like to steal information from web servers because of their widespread use. To prevent being a victim of these scams, it is imperative that you always use a secure browser, particularly when dealing with big sums of money.

Cloud Computing and Its Services

All businesses, no matter how big or little, are gradually adopting cloud services nowadays. The Earth, in other terms, is gradually ascending towards the sky. Since traffic may now flow freely through conventional checks, this new development presents a serious cybersecurity risk. In addition, as the variety of cloud-based software expands, so too must the security measures in place to safeguard sensitive data stored in online apps and cloud services. Even while cloud services are working on their own models, numerous security concerns remain. Despite the benefits the cloud may provide, it's important to remember that as use increases, so do security risks (Nafea, R. A. Et al., 2016).

APT and Targeted Attacks

There is a new generation of malicious programmes out there, and they go by the name "Advanced Persistent Threat" (APT). Web filtering and intrusion prevention systems (IPS) have been important components of network security for years, helping to detect these types of assaults (especially after a first breach). Network security must connect with other security services in order to identify threats as cybercriminals grow more brazen and use more sophisticated methods. So, we need to enhance our security procedures to forestall the emergence of new risks.

Mobile Networks

Modern technology has made it possible to communicate with anybody, wherever in the globe. However, security in modern cellular networks is a major issue. Devices like tablets, smartphones, personal computers, etc., all of which need extra protection beyond what is included in the programmes they "use," have reduced the effectiveness of firewalls and other security measures. The safety of these wireless networks is an ongoing concern. You should exercise extreme caution while using any cellular network that is not your own due to the high risk of becoming a victim of cybercrime.

Ipv6: The New Internet Protocol

IPv6 (Internet Protocol Version 6) is a new internet protocol that will eventually replace IPv4 (Internet Protocol Version 4), the protocol that has served as the internet's backbone since its inception. Safety in IPv6 goes beyond IPv4 compatibility. There are certain basic protocol changes to consider in your security strategy when IPv6 takes over from the bulk issuance of various IP addresses. That's why it's best to make the move to IPv6 as soon as possible: it lowers your vulnerability to cyber-attacks.

Code Encryption

The purpose of encryption is to make it so that unauthorised parties, such as hackers, are unable to decipher encoded communications or data. Using a cryptographic technique, an encrypted message or data is transformed into unintelligible cipher text. As a rule, this is done using an encryption key that dictates the method to be used to encrypt the communication. A data's privacy and security may be safeguarded with only a few easy keystrokes of encryption. However, as encryption's usage spreads, so do the risks associated with it. B. Data transferred across networks (e.g., the Internet, e-commerce), mobile phones, radio microphones, wireless intercoms, etc. are all examples of data that might benefit from encryption while in transit. With the code's encryption, you may check to see whether any sensitive data has been compromised. Logs may be gathered or viewed to provide information about actions taken on the network. The protocol utilised in transmission, as well as incoming and outgoing data packets, are all plainly viewable. Collect data in real time and use an offline analysis sheet to look it over afterwards. Hence the above are some of the trends changing the face of cyber security in the world.

TECHNIQUES OF CYBER SECURITY

Access Control and Password Protection

The combination of a username and a password has shown to be an effective security measure. Perhaps this is your first foray into cyber security.

Data Authentication

All documents we acquire must first be validated, or shown to have originated from a trustworthy source and not been tampered with, before they can be uploaded. Antivirus programmes on the devices verify the authenticity of these files. As a result, reliable anti-virus software is crucial for keeping your devices virus-free.

Malware Scanner

This programme does a complete system scan, examining every file and document for any trace of hazardous code or virus. In this context, the term "malware" is often used to refer to a collection of harmful programmes like viruses, worms, and Trojans (Naseer et al., 2021; Mohamed et al., 2022). Figure 2 depicts the year wise cases for malware.

Figure 2. Amount of malware cases

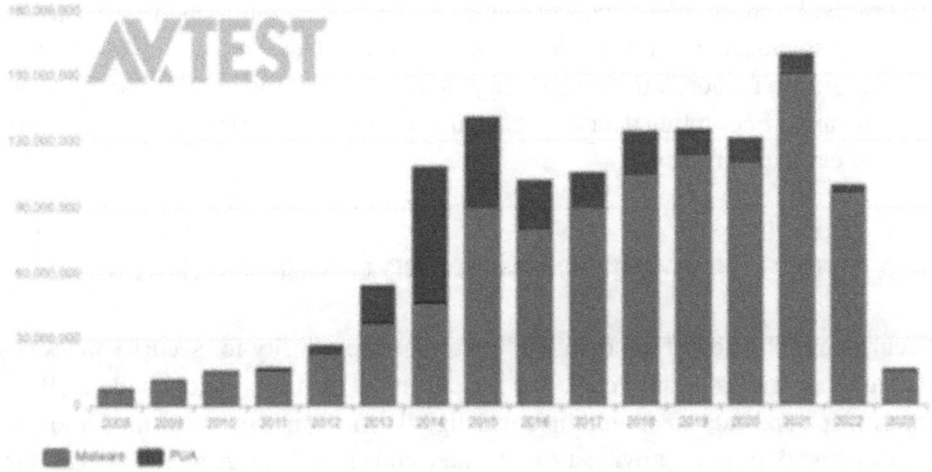

Firewall

A firewall is a piece of software or hardware designed to prevent malicious software like viruses and worms from entering your computer via the network connection you have with the Internet. The existing firewall is responsible for inspecting all incoming and outgoing data transfers to and from the Internet and discarding anything that doesn't pass muster. As a result, firewalls are crucial for protecting against malicious software (Verma, 2022).

Antivirus Software

Antivirus software is designed to identify and eliminate threats from harmful software like viruses and worms. Viruses are constantly being discovered and countered, therefore antivirus software often has an update function that automatically downloads and installs updated virus definitions. A system cannot function properly without antivirus protection.

SOME KEY TECHNOLOGIES TO IMPLEMENT CYBER SECURITY

Access Control and Identity Management: The use of a user name and password to gain access to a computer system has been standard practise since the early 1960s.

Authentication: To ensure that documents are genuine and have not been tampered with, their authenticity must be verified.

Malware Scanners: Malware detection software that performs routine scanning of files and communications.

Firewalls: Firewall software may monitor incoming and outgoing network traffic and sound an alarm if suspicious activity is detected.

Cryptography: There are two primary applications for it in the field of cyber security. Encrypting data at rest and in motion is the most well-known method of ensuring privacy.

STATISTICS OF CYBER CRIMES AND CYBER THREATS

Security and Privacy of the data will always be top priority for security measures that any organization takes care. We are currently living in a globe where all the information and data are maintained in a digital form or in a cyber form. Almost all social networking sites provide a space where end users feel safe themselves as they interact with their friends and family members. Cyber-criminals used to continue to target social media sites to steal personal data and information of home users. It is observed that not only social networking sites but also during banking transactions an users must take all the required security measures (Lu, 2018). Figure 3 shows different types of cyber-attacks.

Figure 3. Different types of cyber-attacks

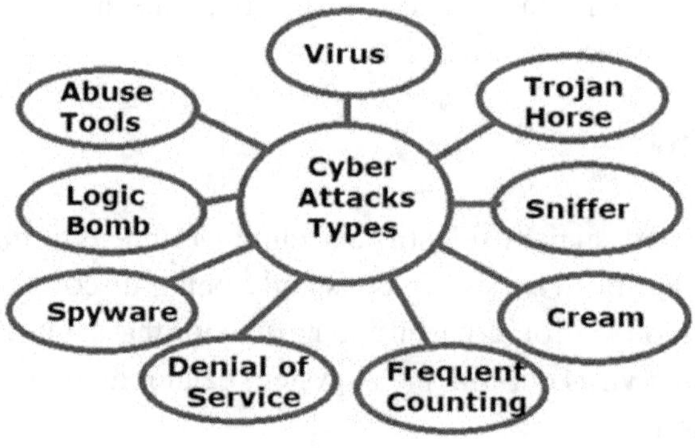

Cyber-crime and cyber threats have grown significantly in recent years, driven by the increasing reliance on technology and the internet in both personal and professional settings.

Figure 4 shows different sources of cyber threats.

Figure 4. Sources of cyber threats.

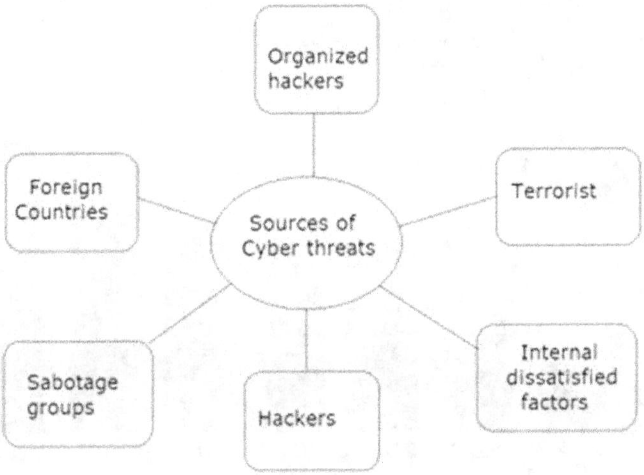

The following are some statistics that highlight the growth in cyber-crime and cyber threats:

- **Increased Frequency:** According to a report from Symantec, the frequency of cyber-attacks increased by 36% from 2019 to 2020. This trend is expected to continue as more individuals and organizations become connected to the internet and become targets for cyber criminals.

According to a report by McAfee, the total number of malware samples reached 1 billion in 2020.The 2020 Cyber Threatscape Report by Accenture found that 60% of cyber-attacks are carried out by external attackers, while 40% are carried out by insiders. The 2020 Cybersecurity Perception Study by Ponemon Institute found that only 5% of organizations believe they are fully prepared to deal with cyber-attacks.

- **Increased Cost:** The cost of cyber-crime is also increasing. A report from Accenture estimates that the global cost of cyber-crime will reach $6 trillion annually by 2021. This represents a significant increase from previous years and highlights the need for organizations to invest in cybersecurity measures to protect themselves from the financial impacts of cyber-crime. Cybersecurity Ventures predicts that cybercrime will cost the world $6 trillion annually by 2021, up from $3 trillion in 2015. Figure 5 represents the estimated cost of cybercrime. Figure 6 represents the cyber security spending respectively.

Figure 5. Estimated cost of cybercrime

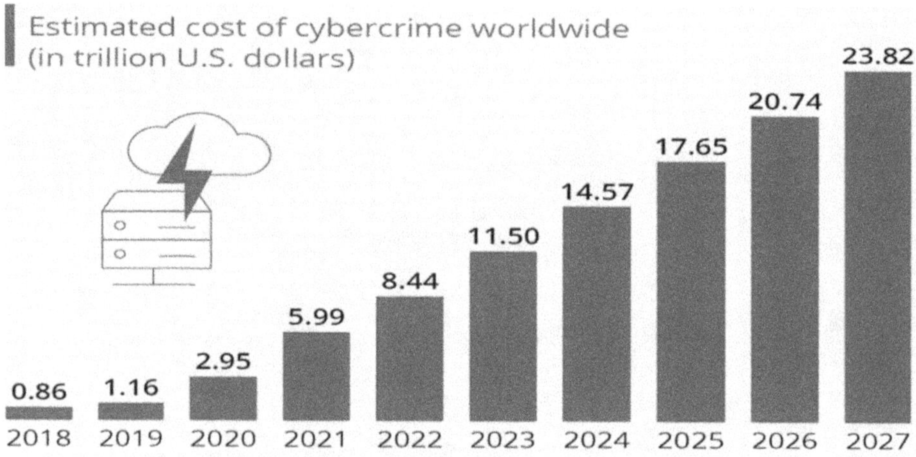

Figure 6. Cyber security spending

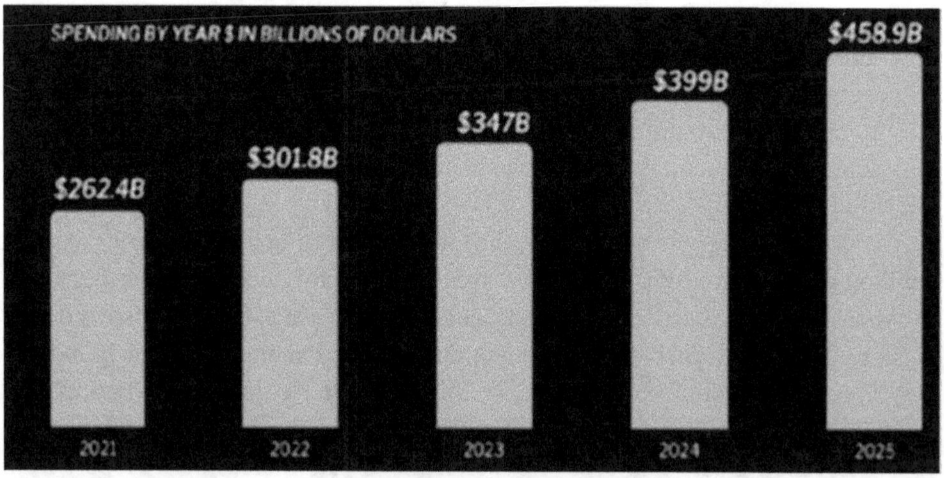

- **Ransomware Attacks:** Ransomware attacks have become one of the most prevalent types of cyber threats. According to a report from Carbon Black, the number of ransomware attacks increased by 41% in 2019. These attacks can cause significant disruption to organizations and result in the loss of sensitive data. The 2020 Cybersecurity Report by Cybersecurity Ventures estimates that a new ransomware attack occurs every 14 seconds (Yamany et al., 2022; Alvee et al. 2021). Figure 7 represents the Global Ransomware attacks by quarterly of recent years.

Figure 7. Global ransomware attacks quarterly

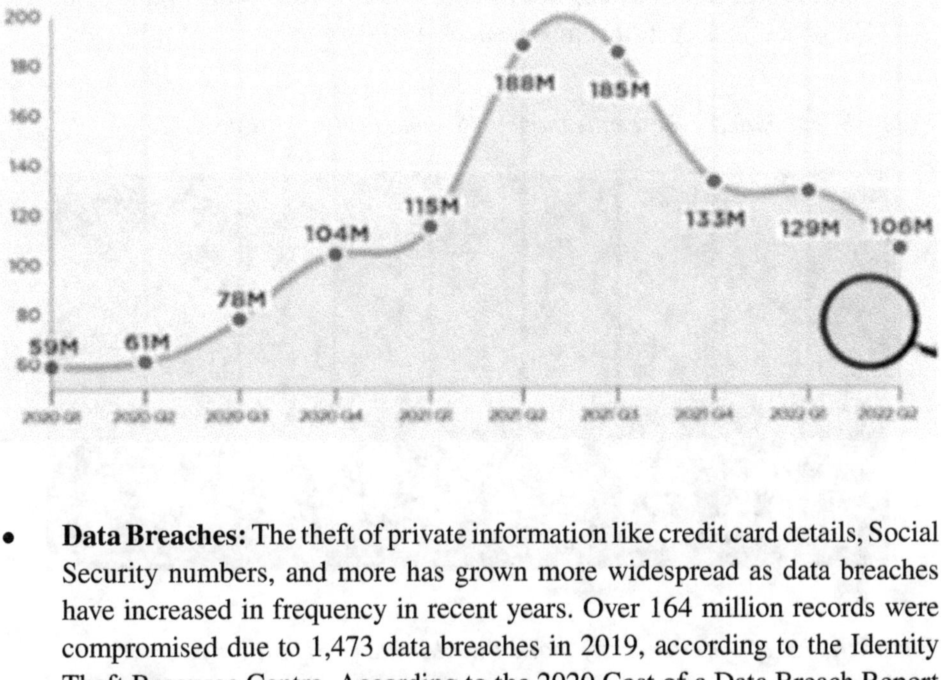

- **Data Breaches:** The theft of private information like credit card details, Social Security numbers, and more has grown more widespread as data breaches have increased in frequency in recent years. Over 164 million records were compromised due to 1,473 data breaches in 2019, according to the Identity Theft Resource Centre. According to the 2020 Cost of a Data Breach Report by IBM, the average total cost of a data breach is $3.86 million. In 2020, the number of data breaches in the United States hit an all-time high of 1,001, up 17% from the previous year, according to the Identity Theft Resource Center (ITRC). The 2020 Data Breach Investigations Report by Verizon found that 86% of data breaches were financially motivated.
- **Phishing Attacks:** The use of phishing assaults, in which hackers send out fake emails or create fake websites to deceive victims into giving over sensitive information, is also on the rise. A report from the Anti-Phishing Working Group found that there was a 32% increase in phishing attacks in the first half of

2020 compared to the same period in the previous year. According to the 2020 State of Password and Authentication Security Behaviors Report by LastPass, 91% of people understand the risk of using the same password across multiple accounts, but 59% still do it anyway. The 2020 Internet Crime Report by the FBI's Internet Crime Complaint Center (IC3) received 791,790 complaints of suspected internet crime, with reported losses exceeding $4.2 billion (Kashif et al., 2018). Figure 8 represents the estimated cybercrime monetary damages in millions USD. Figure 9 shows the size of cyber security market worldwide in billion $. Figure 10 represents the cyber security market size in USD Billion. Figure 11 shows the revenue change in cyber security market by countrywise. Figure 12 represents the cyber security market of our country. All the above figures taken from the resource (Source: Statista Technology Market Outlook, National Cyber Security Organizations, FBI, IMF).

Figure 8. Estimated cybercrime monetary damages (in millions USD)

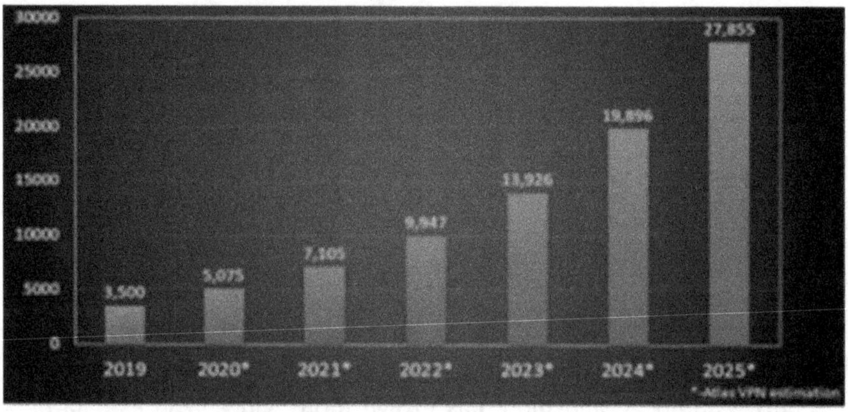

Figure 9. Size of cyber security market worldwide (USD billion)

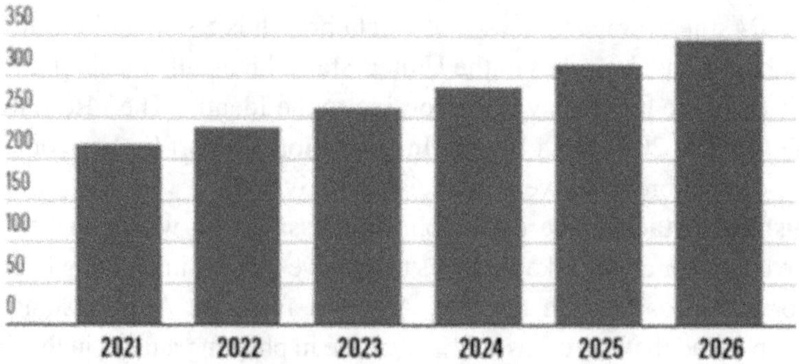

Figure 10. Cyber security market size (USD billion)

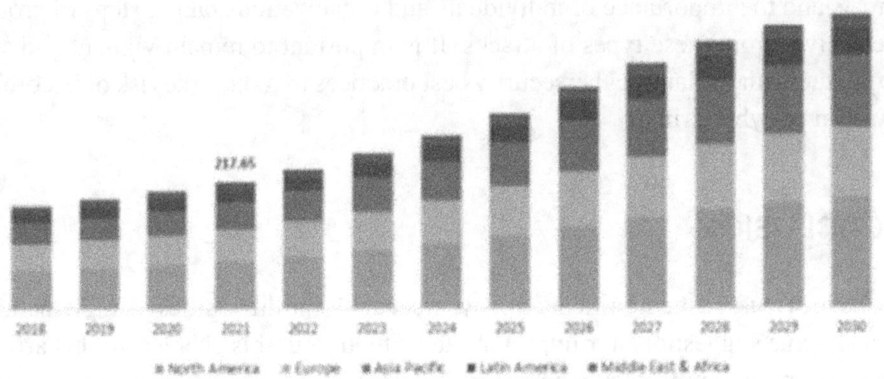

Figure 11. Revenue change in cyber security market by country

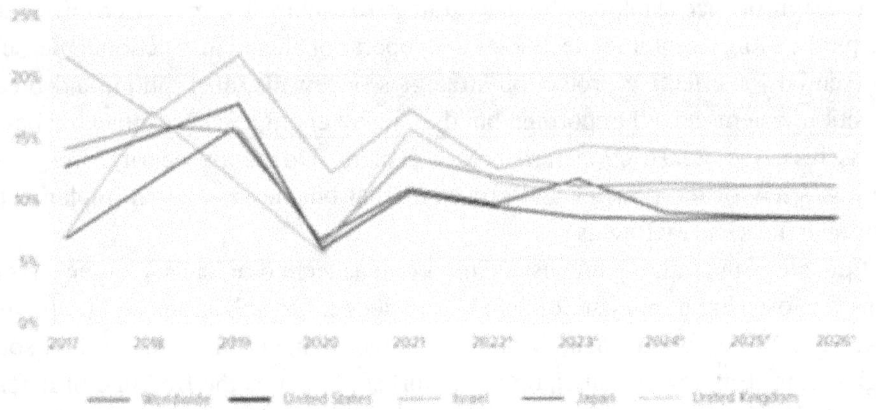

Figure 12. India's cyber security market

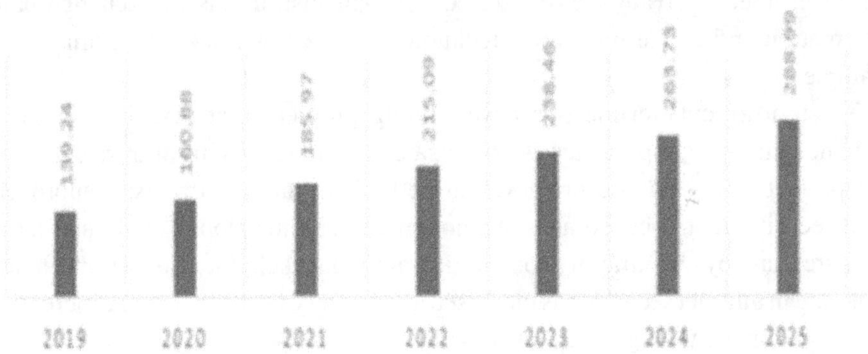

These statistics demonstrate the continued growth in cyber-crime and cyber threats and the importance of individuals and organizations taking steps to protect themselves from these types of attacks. It is important to remain vigilant and stay up to date with the latest cybersecurity best practices to reduce the risk of becoming a victim of cyber-crime.

CONCLUSION

We have discussed the significance of cybersecurity, the difficulties of implementing it, and some suggestions for improving the situation in this chapter. In this article, we've seen how important cybersecurity is in preventing data breaches and other dangers to our digital systems and privacy. Cyber dangers, such as malware assaults and hacking schemes, are becoming increasingly complex, so it is crucial that we keep up with the newest advancements in this area.

The significance of taking a holistic strategy to hacking is emphasised throughout this publication. Combining technological, operational, and instructional measures is essential for effective protection strategies. Firewalls, encryption, and breach detection systems are all important, but they're not enough to keep attacks at bay by themselves. Access restrictions, crisis reaction plans, and routine security assessments are just a few of the policies and processes that businesses should implement to better handle security threats.

Education and training are just as important as technological and organisational steps for fostering a mind-set of safety knowledge. Organizations and individuals alike need to be able to identify and counteract dangers like hacking and social engineering. Employers can equip their workers to act as the first line of defence against cyberattacks by providing them with cybersecurity education and training.

In addition, a multi-stakeholder approach is essential for countering evolving dangers and spreading sound cybersecurity practises. Collaborations between the government, private sector, and educational institutions to exchange danger information and create innovative technologies to counteract cybercrime are one example of this.

To sum up, cybercrime is a never-ending problem that demands continuous vigilance and new approaches. A more safe and robust digital infrastructure, one that protects our data and privacy, can be built by taking a holistic approach to cybersecurity that incorporates technological, organisational, and instructional measures, and by encouraging cooperation among stakeholders. The importance of engaging in internet security has increased as our everyday lives become increasingly dependent on technological advancements.

REFERENCES

Aggarwal, D., & Maurer, U. (2016). Breaking RSA Generically Is Equivalent to Factoring. IEEE Transactions on Information Theory, 62(11), 6251–6259. doi:10.1109/TIT.2016.2594197

Alvee, S. R. B., Ahn, B., Kim, T., Su, Y., Youn, Y.-W., & Ryu, M.-H. (2021). Ransomware Attack Modeling and Artificial Intelligence-Based Ransomware Detection for Digital Substations. In *2021 6th IEEE Workshop on the Electronic Grid (eGRID). 2021 6th IEEE Workshop on the Electronic Grid (eGRID).* IEEE. 10.1109/eGRID52793.2021.9662158

Cremer, F., Sheehan, B., Fortmann, M., Kia, A. N., Mullins, M., Murphy, F., & Materne, S. (2022). Cyber risk and cybersecurity: a systematic review of data availability. In The Geneva Papers on Risk and Insurance - Issues and Practice (Vol. 47, Issue 3, pp. 698–736). Springer Science and Business Media LLC. doi:10.105741288-022-00266-6

Kashif, M., Arshad, S., Tahir, M., Umair, M., & Waqas, P. (2018). A Systematic Review of Cyber Security and Classification of Attacks in Networks. International Journal of Advanced Computer Science and Applications, 9(6). doi:10.14569/IJACSA.2018.090629

Khan, A. W., Zaib, S., Khan, F., Tarimer, I., Seo, J. T., & Shin, J. (2022). Analyzing and Evaluating Critical Cyber Security Challenges Faced by Vendor Organizations in Software Development: SLR Based Approach. In IEEE Access (Vol. 10, pp. 65044–65054). Institute of Electrical and Electronics Engineers (IEEE). doi:10.1109/ACCESS.2022.3179822

Lawnik, M., & Kapczynski, A. (2019). The application of modified Chebyshev polynomials in asymmetric cryptography. In Computer Science (Vol. 20, Issue 3, p. 367). AGHU University of Science and Technology Press. doi:10.7494/csci.2019.20.3.3307

Lee, H., Lee, Y., Lee, K., & Yim, K. (2016). Security Assessment on the Mouse Data using Mouse Loggers. In Advances on Broad-Band Wireless Computing, Communication and Applications (pp. 387–393). Springer International Publishing. doi:10.1007/978-3-319-49106-6_37

Li, J. (2015). The Research and Application of Multi-Firewall Technology in Enterprise Network Security. International Journal of Security and Its Applications, 9(5), 153–162. doi:10.14257/ijsia.2015.9.5.16

Li, Y., & Liu, Q. (2021). A comprehensive review study of cyber-attacks and cyber security; Emerging trends and recent developments. In *Energy Reports* (Vol. 7, pp. 8176–8186). Elsevier BV. doi:10.1016/j.egyr.2021.08.126

Lu, Y. (2018). Cybersecurity Research: A Review of Current Research Topics. Journal of Industrial Integration and Management, 3(4). doi:10.1142/S2424862218500148

Maglaras, L., Janicke, H., & Ferrag, M. A. (2022). Cybersecurity of Critical Infrastructures: Challenges and Solutions. In Sensors (Vol. 22, Issue 14, p. 5105). MDPI AG. doi:10.339022145105

Mathew, A. R. (2019). Cyber Security through Blockchain Technology. International Journal of Engineering and Advanced Technology, 9(1), 3821–3824. doi:10.35940/ijeat.A9836.109119

Mellado, D., Mouratidis, H., & Fernández-Medina, E. (2014). Secure Tropos framework for software product lines requirements engineering. In Computer Standards & Interfaces (Vol. 36, Issue 4, pp. 711–722). Elsevier BV. doi:10.1016/j.csi.2013.12.006

Mohamed, K. F., & Azer, M. A. (2022). Malware Detection Techniques. In *2022 4th Novel Intelligent and Leading Emerging Sciences Conference (NILES). 2022 4th Novel Intelligent and Leading Emerging Sciences Conference (NILES)*. IEEE. 10.1109/NILES56402.2022.9942395

Mohsin, M., Anwar, Z., Zaman, F., & Al-Shaer, E. (2017). IoTChecker: A data-driven framework for security analytics of Internet of Things configurations. In Computers & Security (Vol. 70, pp. 199–223). Elsevier BV. doi:10.1016/j.cose.2017.05.012

Nafea, R. A., & Amin Almaiah, M. (2021). Cyber Security Threats in Cloud: Literature Review. In *2021 International Conference on Information Technology (ICIT). 2021 International Conference on Information Technology (ICIT)*. IEEE. 10.1109/ICIT52682.2021.9491638

Naseer, M., Rusdi, J. F., Shanono, N. M., Salam, S., Muslim, Z. B., Abu, N. A., & Abadi, I. (2021). Malware Detection: Issues and Challenges. Journal of Physics: Conference Series, 1807(1). doi:10.1088/1742-6596/1807/1/012011

Rajasekharaiah, K. M., Dule, C. S., & Sudarshan, E. (2020). Cyber Security Challenges and its Emerging Trends on Latest Technologies. In IOP Conference Series: Materials Science and Engineering (Vol. 981, Issue 2, p. 022062). IOP Publishing. doi:10.1088/1757-899X/981/2/022062

Ramirez, R., & Choucri, N. (2016). Improving Interdisciplinary Communication With Standardized Cyber Security Terminology: A Literature Review. In IEEE Access (Vol. 4, pp. 2216–2243). Institute of Electrical and Electronics Engineers (IEEE). doi:10.1109/ACCESS.2016.2544381

Thakur, K., Qiu, M., Gai, K., & Ali, M. L. (2015). An Investigation on Cyber Security Threats and Security Models. In *2015 IEEE 2nd International Conference on Cyber Security and Cloud Computing. 2015 IEEE 2nd International Conference on Cyber Security and Cloud Computing (CSCloud)*. IEEE. 10.1109/CSCloud.2015.71

Veksler, V. D., Buchler, N., Hoffman, B. E., Cassenti, D. N., Sample, C., & Sugrim, S. (2018). Simulations in Cyber-Security: A Review of Cognitive Modeling of Network Attackers, Defenders, and Users. In Frontiers in Psychology (Vol. 9). Frontiers Media SA. https://doi.org/ doi:10.3389/fpsyg.2018.0069

Verma, A., & Shri, C. (2022). Cyber Security: A Review of Cyber Crimes, Security Challenges and Measures to Control. Vision: The Journal of Business Perspective. doi:10.1177/09722629221074760

Yamany, B., Elsayed, M. S., Jurcut, A. D., Abdelbaki, N., & Azer, M. A. (2022). A New Scheme for Ransomware Classification and Clustering Using Static Features. In Electronics (Vol. 11, Issue 20, p. 3307). MDPI AG. doi:10.3390/electronics11203307

Yim, K. (2010). A New Noise Mingling Approach to Protect the Authentication Password. In *2010 International Conference on Complex, Intelligent and Software Intensive Systems. 2010 International Conference on Complex, Intelligent and Software Intensive Systems (CISIS)*. IEEE. 10.1109/CISIS.2010.185

Chapter 8
Survey and Research Issues on Fog Computing

Rajeshwari Sissodia
Hemvati Nandan Bahuguna Garhwal University, India

ManMohan Singh Rauthan
Hemvati Nandan Bahuguna Garhwal University, India

Varun Barthwal
Hemvati Nandan Bahuguna Garhwal University, India

ABSTRACT

The proliferation of internet-connected devices has led to the accumulation of vast volumes of data, known as "big data." While cloud computing has been effective in managing and processing this data, it falls short in meeting the demands of real-time, low-latency applications and network constraints. To address these limitations, a new computing paradigm called "fog computing" has emerged. Fog computing aims to enhance speed, adaptability, throughput, cryptography, and confidentiality by bringing computation, connection, and memory closer to edge devices and end-users. This chapter provides an overview of the network infrastructure, key technologies, applications, challenges, and unresolved issues associated with fog computing.

INTRODUCTION

Fog computing is a distributed computing paradigm that extends cloud computing to the edge of the network. Fog computing brings computation, storage, and networking resources closer to the end users and devices, enabling faster data processing, lower latency, and improved quality of service.

DOI: 10.4018/978-1-6684-8639-9.ch008

Fog computing is designed to address the limitations of traditional cloud computing, which relies on centralized data centers located far away from end users and devices. This distance can result in delays and network congestion, which can affect the performance of real-time applications such as video streaming, online gaming, and industrial control systems.

Fog computing, on the other hand, brings the cloud closer to the edge of the network, reducing latency and improving the responsiveness of applications. Fog nodes, which are distributed throughout the network, can be deployed in a variety of environments, including factories, hospitals, transportation systems, and smart cities.

Fog computing is also designed to address the challenges of the Internet of Things (IoT), which involves connecting a large number of devices and sensors to the internet. Fog computing can help manage the large volumes of data generated by IoT devices and enable real-time processing and analysis of this data.

LITERATURE SURVEY

Fog computing is an emerging technology that extends the cloud computing paradigm by providing computing, storage, and networking services to the edge of the network. This approach is particularly useful in the context of the Internet of Things (IoT), where billions of devices generate vast amounts of data that need to be processed in real-time.

In recent years, many researchers have investigated various issues related to fog computing, such as security, privacy, resource management, and application development. In this literature survey, we will review some of the most significant contributions in this field.

M. Satyanarayanan et al. (2017) provides an overview of fog computing and its potential applications. The authors discuss the main principles behind fog computing, such as latency reduction, mobility support, and location awareness. They also present several fog computing architectures and use cases, such as smart cities, industrial IoT, and healthcare.

M. H. Rehmani et al. (2017) explores the use of fog computing in smart cities. The authors review the main challenges and solutions in deploying fog computing in smart cities, such as data management, resource allocation, and service provisioning. They also present several use cases for fog computing in smart cities, such as smart traffic management, energy management, and public safety. Finally, they discuss the future research directions in fog computing for smart cities.

H. Zhang et al. (2018) investigates the resource management issues in fog computing. The authors review various resource management techniques, such as load balancing, task scheduling, and energy management. They also analyze several performance metrics, such as response time, throughput, and energy consumption. Finally, they identify some of the challenges in resource management in fog computing, such as resource heterogeneity, mobility, and scalability.

R. Buyya et al. (2018) provides a comprehensive overview of fog computing, including its architecture, enabling technologies, applications, and challenges. The authors review the different layers of the fog computing architecture and highlight the key technologies used in fog computing, such as edge computing, virtualization, and SDN. They also present several application domains for fog computing, such as smart cities, healthcare, and transportation. Finally, they discuss the main challenges in fog computing, such as security, interoperability, and resource management.

S. Aazam et al. (2018) focuses on the security and privacy challenges in fog computing. The authors review the main security and privacy threats in fog computing, such as data breaches, denial-of-service attacks, and identity theft. They also discuss various security and privacy solutions for fog computing, such as encryption, access control, and trust management. Finally, they highlight the future research directions in security and privacy for fog computing.

M. Aazam et al. (2018) focuses on the use of fog computing in healthcare applications. The authors review the main challenges and opportunities in deploying fog computing for healthcare, such as data processing, privacy, and interoperability. They also discuss various solutions for fog computing in healthcare, such as wearable devices, data analytics, and patient monitoring. Finally, they present several use cases for fog computing in healthcare, such as remote diagnosis, telemedicine, and healthcare IoT.

M. Li et al. (2018) explores the use of fog computing in big data analytics. The authors review the main challenges and opportunities in deploying fog computing for big data analytics, such as data processing, data storage, and data security. They also discuss various solutions for fog computing in big data analytics, such as data compression, data caching, and distributed processing. Finally, they present several use cases for fog computing in big data analytics, such as social media analysis, sensor data processing, and financial analysis.

Y. Wang et al. (2018) focuses on the use of fog computing in the context of the Internet of Things (IoT). The authors review the main challenges and opportunities in deploying fog computing for IoT, such as data processing, resource management, and energy efficiency. They also present various use cases for fog computing in IoT, such as smart homes, healthcare, and industrial automation. Finally, they discuss the future research directions in fog computing for IoT.

Z. Zhou et al. (2019) reviews the latest research on fog computing, covering its architecture, key technologies, applications, and open issues. The authors analyze various fog computing architectures, such as centralized, decentralized, and hybrid. They also discuss the main technologies used in fog computing, such as edge computing, virtualization, and machine learning. Finally, they identify several open issues, such as security, privacy, interoperability, and energy efficiency.

N. Kumar et al. (2019) focuses on the security and privacy issues in fog computing. The authors analyze the main threats to fog computing systems, such as data breaches, denial of service attacks, and insider threats. They also discuss various security and privacy mechanisms, such as authentication, access control, and encryption. Finally, they highlight some of the challenges in implementing these mechanisms in fog computing systems, such as resource constraints and heterogeneity.

M. Yousefpour et al. (2019) paper explores the application development issues in fog computing. The authors review various development frameworks and tools for fog computing, such as OpenFog, FogFlow, and Eclipse ioFog. They also analyze the main application domains for fog computing, such as smart cities, transportation, and healthcare. Finally, they identify some of the challenges in application development in fog computing, such as programming complexity, resource limitations, and interoperability.

L. Zhu et al. (2019) focuses on the use of fog computing in multimedia applications. The authors review the main issues and challenges in deploying fog computing for multimedia applications, such as bandwidth constraints, latency, and scalability. They also discuss various solutions for fog computing in multimedia applications, such as content caching, adaptive streaming, and quality of service management. Finally, they present several use cases for fog computing in multimedia applications, such as video surveillance, virtual reality, and augmented reality.

Y. Sun et al. (2019) examines the use of fog computing in 5G networks. The authors review the main challenges and opportunities in deploying fog computing for 5G networks, such as data processing, network slicing, and quality of service management. They also discuss various solutions for fog computing in 5G networks, such as network function virtualization, edge caching, and radio resource management. Finally, they present several use cases for fog computing in 5G networks, such as smart transportation, smart cities, and immersive media.

M. Cao et al. (2019) explores the use of fog computing in smart grid applications. The authors review the main challenges and opportunities in deploying fog computing for smart grid, such as data processing, energy management, and security. They also discuss various solutions for fog computing in smart grid, such as distributed control, data aggregation, and cyber-physical security. Finally, they present several use cases for fog computing in smart grid, such as demand response, renewable energy integration, and power system monitoring.

Z. Chen et al. (2020) explores the use of fog computing in autonomous vehicles. The authors review the main challenges and opportunities in deploying fog computing for autonomous vehicles, such as data processing, communication, and safety. They also discuss various solutions for fog computing in autonomous vehicles, such as edge computing, data fusion, and cybersecurity. Finally, they present several use cases for fog computing in autonomous vehicles, such as intelligent transportation systems, self-driving cars, and unmanned aerial vehicles.

S. Mao et al. (2020) focuses on the use of fog computing in the context of the industrial Internet of Things (IIoT). The authors review the main challenges and opportunities in deploying fog computing for IIoT, such as data processing, energy efficiency, and cybersecurity. They also discuss various solutions for fog computing in IIoT, such as edge intelligence, data analytics, and fault tolerance. Finally, they present several use cases for fog computing in IIoT, such as predictive maintenance, quality control, and smart manufacturing.

L. Zhao et al. (2020) examines the use of fog computing in vehicular networks. The authors review various fog computing architectures for vehicular networks, such as centralized, decentralized, and hybrid. They also discuss the main protocols and technologies used in fog-enabled vehicular networks, such as 5G, edge computing, and V2X communication. Finally, they present several application scenarios for fog computing in vehicular networks, such as intelligent transportation systems, traffic management, and emergency response.

S. Chen et al. (2020) discusses the energy efficiency challenges and opportunities in fog computing. The authors identify several factors that influence the energy consumption in fog computing systems, such as workload characteristics, device heterogeneity, and network topology. They also review various energy-saving techniques, such as task offloading, sleep scheduling, and power management. Finally, they highlight some of the research challenges and future directions in energy-efficient fog computing, such as dynamic resource allocation, energy-aware task scheduling, and green communication technologies.

H. Li et al. (2021) focuses on the use of fog computing in the context of industrial IoT. The authors discuss the main challenges and opportunities in this field, such as real-time processing, data security, and interoperability. They also review various enabling technologies for fog computing in industrial IoT, such as edge computing, wireless communication, and cyber-physical systems. Finally, they present several application scenarios for fog computing in industrial IoT, such as smart factories, intelligent transportation systems, and smart grids.

S. Raza et al. (2021) explores the potential of fog computing in the healthcare domain. The authors review various fog computing applications in healthcare, such as remote patient monitoring, disease diagnosis, and drug delivery. They also discuss the main challenges and opportunities in this field, such as data privacy, security, and interoperability.

FOG COMPUTING ARCHITECTURE

Fog computing architecture is a distributed computing model that brings computing, storage, and networking resources closer to the edge of the network. The architecture consists of several layers that work together to process data and provide services. The following are the main layers of the fog computing architecture as shown in Figure 1.

1. **IoT Devices Layer:** This layer includes a variety of sensors, actuators, and other IoT devices that collect data from the environment. These devices generate a massive amount of data that needs to be processed in real-time.
2. **Edge Layer:** This layer consists of edge devices that are responsible for processing data locally before sending it to the cloud. Edge devices can include routers, switches, gateways, and other network devices that can run applications and services.

Figure 1. Fog computing architecture

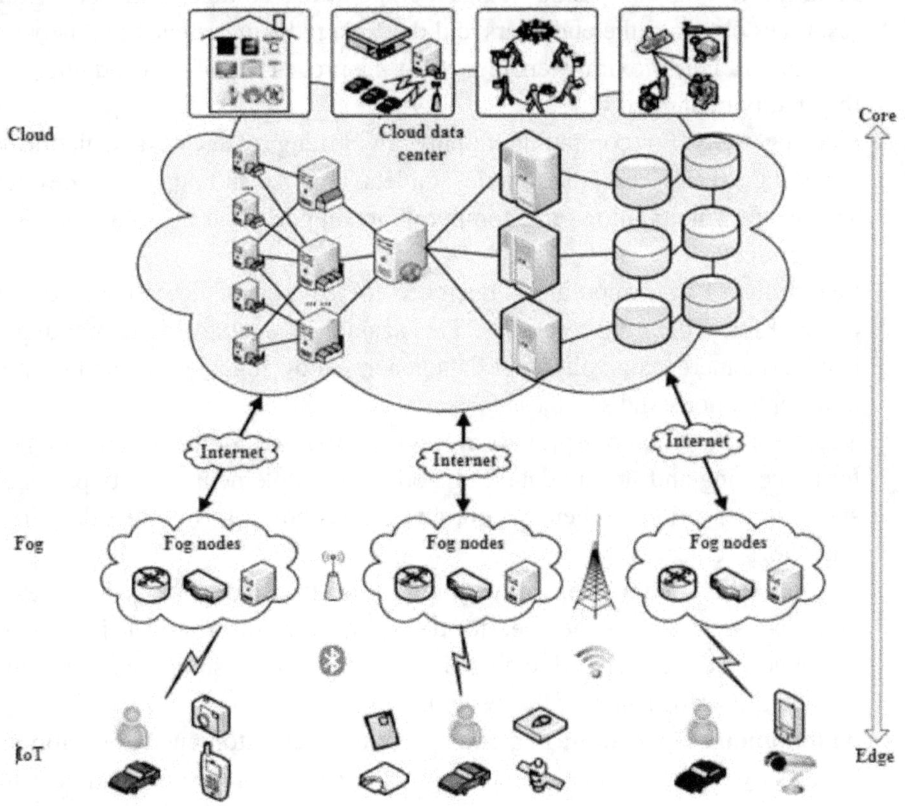

3. **Fog Layer:** This layer includes a set of fog nodes that are deployed closer to the edge devices to process data in real-time. Fog nodes can be deployed in various locations, including on-premise data centers, network edges, and other devices such as routers and switches.
4. **Cloud Layer:** This layer includes a traditional cloud infrastructure that provides services such as storage, computation, and analytics. The cloud layer can process data that is too large or too complex to be processed by the fog layer.
5. **Service Layer:** This layer includes a set of services that are provided to the end-users. These services can include data analytics, machine learning, artificial intelligence, and other services that require significant computing resources.

FOG COMPUTING CHARACTERISTICS

Fog computing is a distributed computing paradigm that extends cloud computing to the edge of the network. The following are the key characteristics of fog computing:

1. **Proximity:** Fog computing brings computation, storage, and networking resources closer to the end-users and devices, reducing latency and network congestion. This proximity enables real-time processing of data and improves the quality of service.
2. **Heterogeneity:** Fog computing supports a wide range of devices and platforms, including IoT devices, smartphones, tablets, laptops, and other edge devices. This heterogeneity allows fog computing to support a variety of applications and services.
3. **Scalability:** Fog computing is designed to scale up or down based on the demand for computing resources. This scalability enables fog computing to handle the increasing volume of data generated by edge devices and support new applications and services.
4. **Security:** Fog computing provides a secure and trusted computing environment for processing and storing data. Fog nodes can implement security protocols such as encryption, authentication, and access control to protect data from unauthorized access.
5. **Resource Efficiency:** Fog computing utilizes local computing resources to process data, reducing the need for data transfer to the cloud. This resource efficiency reduces network congestion and energy consumption, improving the overall performance of the system.
6. **Autonomous Operation:** Fog computing enables autonomous operation of devices by providing local computing resources for processing and decision

making. This autonomous operation improves the responsiveness of devices and reduces the dependence on the cloud for decision making.

DIFFERENCE BETWEEN CLOUD COMPUTING, EDGE COMPUTING, AND FOG COMPUTING

Cloud computing, edge computing, and fog computing are all different models for processing and storing data in modern computing environments. Here are the key differences between them:

1. **Cloud Computing**:
   a. Cloud computing is a model for delivering computing resources, such as servers, storage, databases, and software, over the internet. With cloud computing, users can access these resources on demand and pay only for what they use, without having to invest in expensive hardware and infrastructure.
   b. In cloud computing, all data is stored and processed in remote data centers, which are managed by cloud service providers. Cloud computing offers a high level of scalability, reliability, and security, but it can suffer from high latency and network congestion, especially when dealing with large volumes of data.
2. **Edge Computing**:
   a. Edge computing is a model for processing and storing data at or near the edge of the network, where data is generated or consumed. With edge computing, data is processed locally on devices such as routers, gateways, and sensors, rather than being sent to remote data centers for processing.
   b. Edge computing offers several advantages over cloud computing, including low latency, reduced bandwidth usage, and improved security and privacy. However, edge computing also poses several challenges, including limited computing resources, inconsistent connectivity, and the need for specialized software and hardware.
3. **Fog Computing**:
   a. Fog computing is a model that combines the benefits of cloud computing and edge computing. Fog computing involves distributing computing resources, such as servers, storage, and applications, to the edge of the network, while still maintaining centralized control and management.

With fog computing, data is processed and stored locally on edge devices, but also aggregated and analyzed in a distributed fashion across multiple fog nodes. This approach offers the benefits of edge computing, such as low latency and improved security, as well as the scalability and reliability of cloud computing.

APPLICATION OF FOG COMPUTING

Fog computing has numerous applications across various domains. Here are some examples:

1. **Smart Cities:** Fog computing can be used to manage and control the large number of IoT devices deployed in smart cities. It enables real-time data analysis and processing, allowing for faster response times in emergency situations.
2. **Healthcare:** Fog computing can be used to monitor patients remotely and provide real-time health data analysis. It also enables faster response times in critical situations, reducing the risk of complications.
3. **Industrial Automation:** Fog computing can be used to improve manufacturing processes by providing real-time data analysis and machine learning capabilities. It also enables predictive maintenance, reducing downtime and increasing efficiency.
4. **Transportation:** Fog computing can be used to improve traffic management and reduce congestion. It also enables real-time data analysis for autonomous vehicles, improving safety and reducing the risk of accidents.
5. **Agriculture:** Fog computing can be used to monitor crop growth and soil conditions, enabling farmers to make more informed decisions about irrigation, fertilization, and harvesting.
6. **Retail:** Fog computing can be used to improve customer experiences in retail environments. For example, it can be used to track customer movements and provide personalized recommendations based on their shopping history.
7. **Energy:** Fog computing can be used to optimize energy consumption in buildings by monitoring and controlling heating, cooling, and lighting systems. It can also be used to manage energy grids by providing real-time data analysis and predictive maintenance.
8. **Entertainment:** Fog computing can be used to enable immersive experiences in entertainment venues such as theme parks and theaters. For example, it can be used to synchronize audio and visual effects in real-time, providing a more engaging and interactive experience.
9. **Public Safety:** Fog computing can be used to improve public safety by enabling real-time data analysis and response. For example, it can be used to detect and respond to security threats in public spaces, or to provide emergency services with real-time information in critical situations.

10. **Education:** Fog computing can be used to improve educational experiences by providing real-time feedback and personalized learning experiences. For example, it can be used to monitor student progress and provide targeted interventions to support learning.

CHALLENGES AND OPEN ISSUES OF FOG COMPUTING

While fog computing has many potential benefits, there are also several challenges and open issues that researchers and industry professionals are currently grappling with. Some of the significant challenges and open issues in fog computing includes:

1. **Security and Privacy:** Fog computing introduces new security and privacy challenges, particularly when dealing with sensitive data that needs to be processed and analyzed in real-time. Ensuring the confidentiality, integrity, and availability of data is critical for fog computing applications, and researchers are working on developing new security and privacy mechanisms to address these challenges.
2. **Scalability:** Fog computing requires the ability to scale resources up and down quickly based on demand, which can be challenging. Researchers are investigating various approaches to scaling fog computing infrastructure, such as dynamic resource allocation and load balancing.
3. **Interoperability:** Fog computing involves connecting devices and systems from multiple vendors and platforms, which can result in compatibility issues. Researchers are developing standards and protocols to promote interoperability and ensure that fog computing infrastructure is compatible with various devices and platforms.
4. **Energy Efficiency:** Fog computing infrastructure can consume a significant amount of energy, which can be costly and harmful to the environment. Researchers are exploring various energy-efficient strategies, such as using renewable energy sources and developing low-power hardware and software.
5. **Data Management:** Fog computing generates vast amounts of data that need to be processed, analyzed, and stored in real-time. Researchers are developing new data management techniques, such as data fusion and edge caching, to optimize data processing and reduce latency.
6. **Reliability:** Fog computing infrastructure requires high reliability to ensure that data is processed and analyzed correctly. However, edge devices may experience connectivity issues, and fog nodes may fail, resulting in data loss or delay. Researchers are developing fault-tolerant techniques, such as redundancy and replication, to ensure high reliability in fog computing infrastructure.

7. **Cost:** Building and maintaining fog computing infrastructure can be costly, particularly for smaller organizations or those in developing countries. Researchers are exploring cost-effective solutions, such as using low-cost hardware and open-source software, to reduce the overall cost of deploying and maintaining fog computing infrastructure.
8. **Governance and Regulation:** Fog computing introduces new governance and regulatory challenges, particularly with respect to data privacy, security, and ownership. Researchers are examining various governance models and regulatory frameworks to ensure that fog computing infrastructure complies with legal and ethical standards.
9. **Human Factors:** Fog computing involves connecting devices and systems to a distributed computing infrastructure, which can be challenging for non-technical users. Researchers are investigating human factors issues, such as user interfaces and training, to ensure that fog computing infrastructure is user-friendly and accessible to all users.
10. **Performance Optimization:** Fog computing requires optimizing the performance of the computing infrastructure to ensure that it meets the demands of real-time applications. Researchers are developing performance optimization techniques, such as task scheduling and resource management, to ensure that fog computing infrastructure can handle diverse and dynamic workloads.

CONCLUSION

In conclusion, the increasing usage of internet-connected devices and the generation of large amounts of data have led to the development of fog computing. This new computing paradigm aims to complement cloud computing by bringing computation, connection, and memory closer to edge devices and end-users, thereby improving speed, adaptability, throughput, cryptography, and confidentiality. This literature survey has provided an overview of the network infrastructure, essential technologies, applications, challenges, and unresolved issues of fog computing. It has also discussed the similarities and differences between fog, cloud, and edge computing, and presented various challenges that need to be addressed to advance the subject of fog computing. Overall, fog computing is a promising approach that has the potential to revolutionize how we process and store data, especially in real-time low-latency applications. However, there are still many unresolved issues that need to be addressed to fully realize the potential of fog computing. Further research and development are needed to address these challenges and advance the subject of fog computing.

REFERENCES

Aazam, M. (2018). Fog computing for healthcare: A review of current trends and future directions. *Journal of Medical Systems*, *42*(7), 1–13.

Aazam, S. (2018). Fog computing: Issues and challenges in security and privacy. *Journal of Network and Computer Applications*, *116*, 1–15.

Buyya, R. (2018). Fog computing: A survey of enabling technologies, applications, and issues. *IEEE Internet of Things Journal*, *5*(2), 524–545.

Cao, M. (2019). Fog computing for smart grid: A survey. *IEEE Transactions on Industrial Informatics*, *15*(1), 104–118.

Chen, S. (2020). Energy-efficient fog computing: A review of research challenges and future directions. *IEEE Network*, *34*(2), 22–28.

Chen, S., Yang, H., Gao, X., Lin, W., & Zhang, H. (2020). Energy-efficient fog computing: A review of research challenges and future directions. *Journal of Parallel and Distributed Computing*, *140*, 99–112.

Kumar, N. (2019). Fog computing: Issues and challenges in security and privacy. *Journal of Ambient Intelligence and Humanized Computing*, *10*(3), 1183–1198.

Li, H. (2021). Fog computing for industrial IoT: A review of enabling technologies and applications. *IEEE Internet of Things Journal*, *8*(9), 7426–7442.

Li, H., Yang, H., Zhou, L., & Jiang, C. (2021). Fog computing for industrial IoT: A review of enabling technologies and applications. *Journal of Ambient Intelligence and Humanized Computing*, *12*(2), 2109–2122.

Li, M. (2018). Fog computing for big data analytics: A survey. *Journal of Network and Computer Applications*, *98*, 18–28.

Raza, S. (2021). Fog computing for healthcare: A review of applications, challenges, and future directions. *IEEE Journal of Biomedical and Health Informatics*, *25*(6), 1876–1893.

Raza, S., Shafi, S., Saeed, U., & Imran, M. (2021). Fog computing for healthcare: A review of applications, challenges, and future directions. *IEEE Access : Practical Innovations, Open Solutions*, *9*, 52363–52381.

Rehmani, M. H. (2017). Fog computing for smart cities: A review of challenges and solutions. *Journal of Ambient Intelligence and Humanized Computing*, *8*(1), 127–142.

Satyanarayanan, M. (2017). Fog computing: Principles, architectures, and applications. In *Proceedings of the 2017 Workshop on Fog Computing and the IoT* (pp. 23-27). Academic Press.

Wang, Y. (2018). Fog computing for the Internet of Things: A survey. *IEEE Internet of Things Journal*, *5*(2), 854–867.

Yousefpour, M. (2019). Application development in fog computing: A survey. *IEEE Access : Practical Innovations, Open Solutions*, *7*, 112792–112809.

Zhang, H. (2018). Resource management in fog computing: A survey. *Journal of Network and Computer Applications*, *103*, 1–14. doi:10.1016/j.jnca.2018.08.013

Zhao, L. (2020). Fog-enabled vehicular networks: A review of architectures, protocols, and applications. *IEEE Communications Surveys and Tutorials*, *22*(1), 559–584.

Zhou, Z. (2019). A survey on fog computing: Architecture, key technologies, applications, and open issues. *IEEE Communications Surveys and Tutorials*, *21*(2), 1206–1258.

Zhu, L. (2019). Fog computing for multimedia applications: A review of issues and challenges. *IEEE Communications Surveys and Tutorials*, *21*(4), 3385–3408.

Chapter 9
Virtual Reality, Augmented Reality, and Video Games as Important Factors in Health and Education

Marva Angélica Mora Lumbreras
https://orcid.org/0000-0001-6505-2205
Universidad Autónoma de Tlaxcala, Mexico

ABSTRACT

Some areas of priority attention in Mexico are health and education, which open an opportunity for technology, specifically this chapter presents different projects that use augmented reality (AR), virtual reality (VR), or video games (VG), which provide an excellent contribution to education and the health for its dynamism and interactivity, these projects were developed in the faculty of basic sciences, engineering, and technology (FCBIyT). The projects presented for medicine using AR are Arachnophobia, TEAPRENDE, and SARA (food labeling), AVHI: Non-Hodkin Lymphoma uses VR, and Byoky, a video game. The projects developed for education with AR are LenSeñasMx and Industrial Revolution, the projects that uses VR are the diasporas of Tlaxcala, technological advances by James Watt and the building of a giant virtual computer with this multi-screen environment. Video games for education are Independence of Mexico and the diasporas of Tlaxcala too.

DOI: 10.4018/978-1-6684-8639-9.ch009

INTRODUCTION

According to CONACYT's National Strategic Programs, "health" and "education" are priorities in Mexico (Strategic National Programs, n.d.), this chapter presents different projects carried out at the FCBIYT focus on health and education.

In health, AR has its area of opportunity, for example, in phobias, as arachnophobia, which allows the psychologist to treat the patient gradually, the patient will not have to imagine the spider, but he could see a small spider on the floor through a mobile device. Even, AR can be useful for autism spectrum disorder (ASD), where some habitual, social and educational behaviors are shown to the child with ASD. On the other hand, currently with the advance of diabetes in Mexico, on October 1, 2020, the Official Mexican Standard NOM 051 began to manage food labeling (gob.mx, 2021), for which it was decided to give an explanation using AR of what happens in the body when there is excess sugar, sodium, calories, saturated fat, trans fat. So, when a person decides to buy a product and has different labels when using the application installed on their cell phone, they will be able to know how the product they plan to buy could affect the body, giving them the possibility of limiting or refraining from consuming it.

VR can be an important ally in health, for example, in cancer, since through software the user can be presented with what is happening in the body of a Non-Hodkin Lymphoma patient, it can even be used of a haptic interface to give it greater realism and get to know a certain texture of the organs.

A VG can also contribute to the area of health, for example, a video game that presents a story of a patient with dissociative identity disorder or multiple personality disorder, the video game could show how this disease develops, so their relatives could know that is happening in the patient, which will make them aware of the subject, in a certain way to know what could happen to their family member.

On the other hand, it can be said that AR, VR and VG also have their contribution in education, culture and traditions, some works successfully developed under my direction are described below.

AR could be very useful to complement the teaching of the Mexican Sign Language (LSM), making sections of words such as colors, numbers, school supplies, furniture, among others, which is supported by a marker, indicate how the hands move. to say something, giving the opportunity for those interested in learning LSM to rely on this type of application. Work has also been done with inventions developed in the Industrial Revolution using AR, giving the user the ability to see some inventions in 3D and at the same time listen to an explanation of them, all from a mobile application.

VR is being useful in education, when using it as a complement to what is taught at school, only in the history area could we find software that shows historical aspects in a dynamic way, for example, having a virtual tour of the diasporas of Tlaxcala, indicating that the Viceroy Don Luis de Velasco II in March 1591 recruited 400 Tlaxcalan families to found towns in the northern New Hispanic zone, an area threatened by the Chichimeca War, at the same time that it is shown the contributions that the Tlaxcalans took during the diaspora. In the same way with VR you can take a virtual tour of technological advances by James Watt, installed throughout a small town, showing them in 3D with an explanation of their operation. VR is not only useful in history, how about building a computer viewed through multiple surround screens, using your hands to build a giant virtual computer, with this multi-screen environment the student will learn relevant parts of the computer to function properly.

A VG also allows you to tell the history, for example, important events of the Independence of Mexico, in which Spanish rule ended. VG with 3D characters, but with 2D movements, showing such important details as the participation of Miguel Hidalgo, Josefa Ortiz de Domínguez, José María Morelos and Pavón, among others. Or how about learning about the diasporas of Tlaxcala, but now with a VG in which the user learns about the history through challenges and activities.

As mentioned before, AR, VR y and VG are suitable for Health and Education, allowing users of different ages to take advantage of them through these technologies.

AUGMENTED REALITY, VIRTUAL REALITY, AND VIDEO GAMES

Augmented Reality is the combination of the real world with information such as text, audio, video, 2D and/or 3D graphics, in real time and from a mobile device, it can be using markers, which are currently QR, photos, drawings, text or geopositioning, which is to locate yourself at a cartographic point.

Virtual Reality is a virtual environment where the user is immersed in a computer-generated world, using the human senses through different devices, to mention some: sensors, cameras, haptic controls, virtual reality glasses, among others. There are different levels of immersion, such as a multi-screen world, Virtual Reality glasses, or worlds on a computer or cell phone, which are the lowest cost.

A video game is a game that can be played in a software or application, which gives the player interaction to do challenges and activities in it.

PROJECTS FOCUSED ON HEALTH

To start, it can be said that Augmented Reality, Virtual Reality and Video Games are very useful in health, some works developed with some students with positive results during their tests are mentioned below.

Arachnophobia

A phobia is the reaction of great fear towards something. In this way, the term Arachnophobia is defined as the persistent, abnormal and unjustified emotion of fear of spiders (Bourdon et al., 1988). Since the 90's, technology has been incorporated into the field of treating phobias and other psychological disorders. If different virtual tools can provoke fear (Rothbaum et al., 2001), exposure through Augmented Reality can work as an alternative mode of inducing exposure.

A project called Arachnophobia is proposed to help the psychologist to treat the fear of spiders gradually. Arachnophobia is a mobile application based on the Android platform, which provides a series of activities for the visualization of phobic environments (specifically based on Arachnophobia) through Augmented Reality, which allows the user to interact with a three-dimensional world generated through mobile devices. The system is composed of the following modules:

A. **Test** is a guide to indicate how much fear a user feels towards spiders. It is worth mentioning that the questionnaire does not provide a diagnosis of spider phobia, nor is it a substitute for a diagnosis that would be given by an expert in the area.
B. **Aractnophobia** presents the definition of Arachnophobia.
C. **Gallery** presents anaglyphic photographs of spiders.
D. **Augmented Reality Simulation** shows the simulation of a spider with Augmented Reality, this module allows the psychologist to gradually treat the patient and show a spider in different sizes and positions in the doctor's office
E. Credits shows project participants

Figure 1 shows an anaglyphic photograph of a spider of the section gallery.

This project was presented and tested by a group of 20 people, during the tests most of the people were able to install Arachnophobia without problem.

- Navigability. 70% of people had not difficulty navigating in the application because pressing some buttons. The bug was fixed in the final version of the app.
- Interface. Most of the people liked the interface and it seemed correct and according to the topic that is being discussed. 20% of people suggested some changes regarding color, button placement, images, etc.

Figure 1. Anaglyphic photograph of a spider

- Content. They found the content interesting, they commented that the survey seemed very useful, as well as the information about spiders, especially the augmented reality part.
- Innovation. The app seemed very innovative and interesting to most of the people.

The application was very well accepted by the users, this is because the application seemed innovative and met the aforementioned objectives. Some comments from users were considered and the observed corrections were made, which were mostly related to the interface and navigability.

Therefore, it can be concluded that the results of the surveys on the application were satisfactory and that they largely meet the requirements of the application.

At present there are a large number of works related to the topic of augmented reality, however, it has been noted that in Mexico no projects related to phobias have been developed, this was one of the reasons why it was decided to carry out this project and it was chosen to address the topic of Arachnophobia.

TEAPRENDE

Autism spectrum disorder (ASD) is a developmental disability caused by differences in the brain. People with ASD often have problems with social communication, interaction, restricted or repetitive behaviors. People with ASD may also have different ways of learning, moving, or putting attention. It is important to note that some people without ASD might also have some of these symptoms. But for people with ASD, these characteristics can make life very challenging (ASD, 2023).

Considering that the ASD can be positively supported with technology, the idea of programming the TEAPRENDE project arises. TEAPRENDE is an Augmented Reality application focused on autism spectrum disorder (ASD), this application is developed for children from 5 to 9 years old, the application is used with Android mobile phones, and it has the following modules:

A. **Recognition and Expression of Emotions**. This module contains the emotions which are: joy, disgust, anger, sadness, fear and amazement, which were modeled in 3D using Blender.
B. **Learn the Letters**. This module contains the letters of the alphabet, and an 3D object that begins with the letter presented.
C. **Learn Numbers**. This module lets to learn numbers with striking 3D objects that capture the attention of the child.
D. **Learn Geometric Bodies.** This module presents the following geometric bodies: Sphere, Pyramid, Cylinder, Pentagonal Prism, Rectangular Prism and Cube.
E. **Repeat Basic Habits.** It is known that children with ASD have a difficult doing daily activities at home or anywhere else, so in this module the children learn to do things and remember them, such as bathing and brushing their teeth.
F. **Learn Social Norms.** Social skills are considered a set of abilities that allow us to develop a repertoire of actions and behaviors that make people function in the social sphere. Children often fail in social competence; It is difficult for them to be able to use the different visual cues, these being fundamental in the social contexts in which they operate, especially at school. This module introduces, greetings and dialogues, see Figure 2.

On April 26, 2022, the work team went to CAM 15 located in Contla de Juan Cuamatzi to request permission to test the TEA application in 3 groups of children with Autism. After requesting the pertinent permits from principal and parents, the tests were carried out on students in an age range of 5 to 8 years. The results obtained in the tests are described below:

- **Recognition and Expression of Emotions.** The test consisted in that the children knew the section of recognition and expression of emotions through the application the different emotions were shown in 3D, such as: joy, disgust, anger, sadness, fear and amazement.
- **Learn 3D Geometric Bodies.** It consisted in that through the application the geometric bodies in 3D were presented: sphere, pyramid, cylinder, pentagonal prism and rectangular prism.

Figure 2. Learn social norms dialogue

- **Learn the Letters of the Alphabet.** In learning letters of the alphabet, different letters were shown to the children, each one appears with an object or animal that begins with that letter, as a basic exercise for children to learn letters and objects, as well as identify them in their environment.
- **Repeat Basic Habits.** The repetition of basic habits is important for the children, who benefit from the constant use that helps them to know how their daily activities should be.
- **Learn Basic Numbers.** Learning basic numbers is essential, since they encourage learning intuitively. All these stimulate creativity, emotion, reasoning and calculation, since this shows the numbers from 1-10 with the number of corresponding objects.
- **Learn Social Norms.** Social skills are a set of habits or skills that help the child to relate to the environment. The objective of the module seeks to reinforce basic social norms through animations.

The children's response was adequate, it was possible to appreciate how they paid attention and got excited with the animations included, it is worth mentioning that in order to carry out the tests, they had the support of the teachers, with whom the children are familiar. All the materials presented are 3D.

SARA

In Mexico, the labeling of prepackaged food and non-alcoholic beverages of national or foreign manufacture, are labelled, is mandatory in accordance with NOM-051-SCFI/SSA1-2010, which is in charge of establishing commercial information. and health that these products must contain. In addition, NOM 051 determines the characteristics of said information and establishes a frontal labeling system, which must warn the final consumer clearly and truthfully about the content of critical nutrients and ingredients that represent a health risk when consumed in excess. This labeling is made up of 5 octagon-shaped warning seals, which clearly, simply and visibly indicate when a product contains excess nutrients and critical ingredients such as: calories, saturated fats, trans fats, sugar and sodium (gob.mx, 2021).

Due to it is important for the consumer to know what food stamps mean, the idea of programming an application that makes a brief and precise explanation of what happens when consuming the excess ingredients that each food product has emerged.

SARA (Augmented Reality focused on Food Stamps) is an application that support society in a new way. It allows through AR to explain to the consumer what will happen to the body when they consume either the packaged food or the drink. The application has the options: Scan, Gallery, About and Contact, see Figure 3.

Figure 3. SARA augmented reality focused on food stamps

A. **Scan** allows you to scan the stamps on any product that contains them.
B. **Gallery** shows a menu that has 5 options that represent each stamp: excess sodium, excess sugar, excess trans fat, excess saturated fat and excess calories, each one has its own gallery of products and consequences for consuming it.
C. **About** shows the purpose for which the application was developed
D. **Contact** allows the visualization of the developer's contacts

For this project, modular and integration tests were carried out. Integration tests are those that are carried out in the field of software development once the modular tests have been approved and what they prove is that all the modular elements that make up the software work correctly by testing them as a group.

In addition, usability tests were carried out that measure the ease of use of a software. For the tests of the SARA application, the mobile devices with the Android operating system that belonged to the test users. The test users were 54 people between the ages of 15 and 45, some with prior knowledge of food stamps and others with no prior knowledge of how the stamps work.

When conducting the test, it was obtained that the application developed that shows information regarding what happens in the body when they consume excess of an ingredient marked with the food stamp, had a great acceptance, because in general people do take the information into account, so you are aware that every bite you eat could have serious consequences.

AVHI

Non-Hodgkin lymphoma is a disease in which malignant (cancer) cells form in the lymph system. Non-Hodgkin lymphoma can be indolent or aggressive. Older age, being male, and having a weakened immune system can increase the risk of adult non-Hodgkin lymphoma. Signs and symptoms of adult non-Hodgkin lymphoma include swollen lymph nodes, fever, drenching night sweats, weight loss, and fatigue. Tests that examine the lymph system and other parts of the body are used to diagnose and stage adult non-Hodgkin lymphoma (NCI, 2022).

Non-Hodgkin lymphoma can be explained in different ways, one of them is with technology, specifically Virtual Reality can help, given its dynamism and interactivity.

AVHI: Non-Hodkin Lymphoma (NHL) uses VR, it is software developed on Windows platform. It includes four main aspects: definition of the disease, main organs of the lymphatic system, diagnosis and possible treatment. The user interacts with the environment through a haptic device and a PC keyboard.

A. **Definition of the Disease**. The first section of the AVHI is the definition of Non-Hodkin Lymphoma disease.

B. **Main Organs of the Lymphatic System**. In the AVHI, four main organs of the lymphatic system were included: the thymus, the spleen, the bone marrow and the tonsils.

C. **Symptoms**. The main symptoms are swollen lymph nodes in the abdomen, groin, neck or armpits. Swollen liver or spleen. Recurring fever not related to an infection or other illness, weight loss with no known cause, sweating, chills and fatigue.

D. **Diagnosis.** Explains that the diagnosis can be detected through different tests such as: biopsy (tissue removal), computed tomography (3D image of the inside of the body), magnetic resonance imaging (detailed images of the body) and positron emission tomography (images of internal organs and tissues).

E. **Treatment**. Explains that there is four main treatments for NHL: chemotherapy, radiation therapy, immunotherapy, and targeted therapy.

F. **Rules Against Cancer.** It presents the rules suggested by the Mexican Association for Cancer Control, A.C. are: not smoking, eating fresh fruits and vegetables, avoiding excess weight, avoiding prolonged exposure to the sun without protection, do not expose yourself to substances considered carcinogenic, see a doctor if there is an abnormal growth, a wound that does not heal, a mole that changes (in shape, size or color), or any abnormal blood loss, see a doctor in case of permanent hoarseness, change in bowel habits, urinary disturbances, or abnormal weight loss, women should have a Pap smear once a year, self-examination, vaginal ultrasound, and mammography. Men should regularly examine testicles, after age 40, get vaccinated against human papillomavirus and hepatitis B.

Figure 4 presents the section "Main organs of the lymphatic system", specifically the bone marrow.

Figure 4. Organs of the lymphatic system: Bone marrow

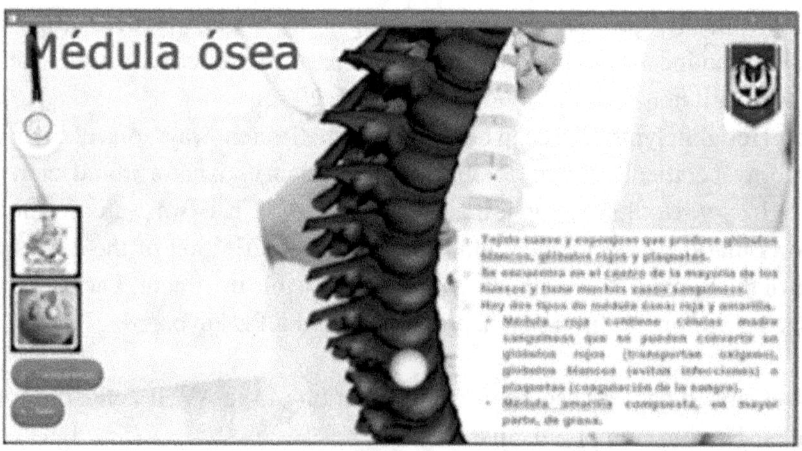

To carry out the evaluation of the project, a usability test was designed that contains different sections where presentation, trust and credibility are evaluated. To carry out the tests, the software was shown to 27 people, whose ages were between 19 and 23 years old, of whom 44% were women and 55% were men. The test was carried out in two sessions, in the first some details could be refined, and the results were improved in the second evaluation, the results are described below.

First, the presentation of the interface was evaluated. In evaluation 1, when the respondents were asked about the general functioning of the AVHI, 70% of positive results were obtained, in the second evaluation, after the feedback and corrections, 94% of positive results were obtained. When inquiring if the AVHI design is considered professional in the first evaluation, 67% was obtained and 82% in the second test. To verify if the AVHI design encourages users to explore it more, 73% were obtained in the first evaluation and 82% in the second. For the evaluation of credibility and trust, when verifying whether the user considers the AVHI reliable, 80% positive results were obtained in the first evaluation and 88% in the second evaluation. When evaluating the visual design of the VA, 70% of positive results were obtained in the first evaluation and 88% in the second. After seeing the results, it can be said that AVHI is interesting to the general public and could help in a relevant way to provide information related to non-Hodgkin lymphoma.

Byoky

Dissociative disorders are mental disorders that involve experiencing a disconnection and lack of continuity between thoughts, memories, surroundings, actions and identity. People with dissociative disorders escape reality in ways that are involuntary and unhealthy and cause problems with functioning in everyday life. Dissociative disorders usually develop as a reaction to trauma and help keep difficult memories at bay. Symptoms — ranging from amnesia to alternate identities — depend in part on the type of dissociative disorder you have. Times of stress can temporarily worsen symptoms, making them more obvious (Mayo Clinic, n.d.).

Capgras syndrome (CS), or delusion of doubles, is a delusional misidentification syndrome. It is a syndrome characterized by a false belief that an identical duplicate has replaced someone significant to the patient. In CS, the imposter can also replace an inanimate object or an animal. Contrary to the earlier belief that CS mainly affects women, it occurs in both genders. It is widely regarded as the most prevalent of the delusional misidentification syndromes and appears in psychiatric and non-psychiatric cases, including patients with brain damage (NLM, 2022).

Dissociative identity disorder and Capgras syndrome can be very complex diseases, which are not entirely common, however, it is important to know about them, as well as their symptoms, firstly for family members, but also for the general public. Thus, analyzing the importance of these diseases, the video game Byoky is developed at the FCBIyT.

Byoky is a 2D video game focused on two mental disorders: Dissociative Identity Disorder and Capgras Syndrome. It has a series of events of animations, dialogues and music. Through the video game, the player will select the option that seems best to him according to the situation in which he finds himself. There are key rooms where there are pieces that will make it possible to assemble the best memory of the star character of the level and will open the door to the next level. Throughout the game there will also be characters that will make the journey not so easy, these characters will graphically represent the disease of level.

A. **The First Level Is Called Dramaturgy**, through the different rooms the player will see the behavior of the main character to understand the Dissociative Identity Disorder.
B. **The Second Level Called Double Bind**, the main character suffers from capgras syndrome. In order to progress, the player must understand the story of the main character and the condition that is presented in this level, so the interaction with said main character will take a more active role in the story, since they must also understand why of the syndrome.

Figure 5 shows a map of Byoky. Byoky is a video game 2D developed in RPG Maker.

Figure 5. Byoky

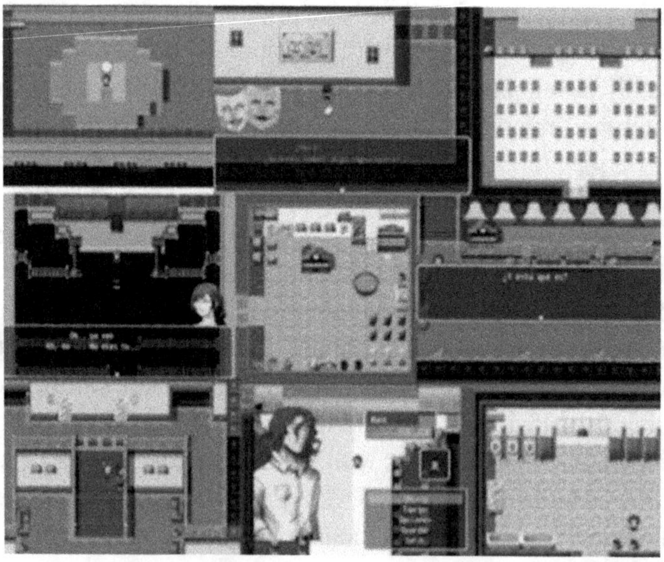

This project was presented to the psychologists of the Faculty of Basic Sciences, Engineering and Technology, who considered that Byoki is pertinent so that relatives and people of the general public interested in the subject could learn about both diseases. In addition, Byoki: Dissociative Identity Disorder and Capgras Syndrome Video Game, was tested by 21 people who evaluated trust and credibility as 80% positive.

LenSeñasMx

Mexican Sign Language (LSM) is the majority language of the deaf community in Mexico. Previously, a number between 87,000 to 100,000 signers was estimated (Smith-Stark, 1986), now, based on the very high number of deaf people from birth and prior to old age in the 2010 and 2020 censuses, it is estimated population using it to be approximately 300,000 (Faurot et al., 1992).

Considering the great growth that has occurred lately in the deaf community, LenSeñasMx is developed for family members and the general public who want to learn about Sign Language.

LenSeñasMx is an application developed with Augmented Reality focused on the mexican sign language (LSM) and is developed for Android devices. This project was carried out with the support of the deaf community of Apizaco. LenSeñasMx is divided in four categories:

A. School supplies
B. Colors
C. Verbs
D. Numbers.

Figure 6 shows the LenSeñasMx application, the boy's sign represents the number 7 in LSM.

LenSeñasMx was tested by modules and by application integration tests. As well as usability tests were carried out on 20 users whose age is between 18 and 25. The usability test consisted of 10 questions, in which the user's experience when using the application is sought, in addition to evaluating the ease of the application, its quality and the use cases proposed. Before applying the survey, a brief description of the application was given to the users, after which they were allowed some time to familiarize themselves with the application. Once the above was completed, the survey was applied, whose rating could be between 0 and 10. The Presentation of the application was evaluated with 9.4, Task orientation obtained a 9.2 Content and writing obtained 9.1, Navigability obtained 9.1. Giving a total 9.2 overall rating.

Figure 6. LenSeñasMx presents the representation of number 7 in LSM

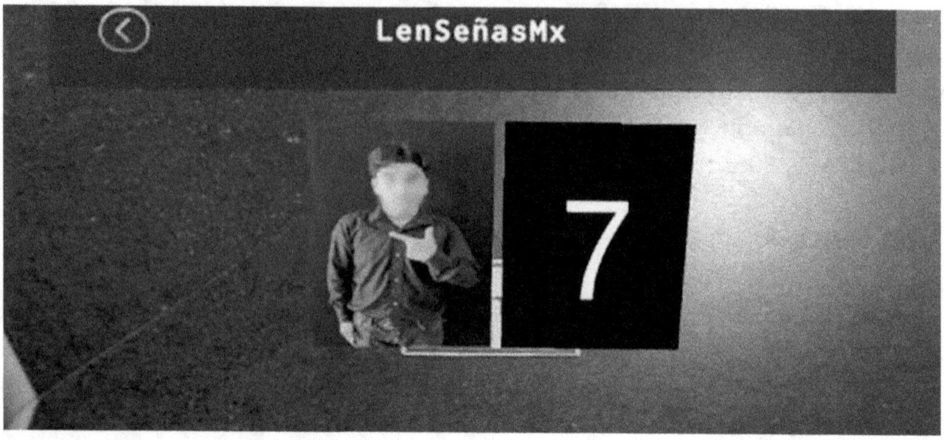

PROJECTS FOCUSED ON EDUCATION

Below are various projects developed either with Augmented Reality, Virtual Reality or Video Games, useful for education. These projects have been tested with modular, integration and usability tests.

Inventions of the Industrial Revolution in AR

Industrial Revolution, in modern history, the process of change from an agrarian and handicraft economy to one dominated by industry and machine manufacturing. These technological changes introduced novel ways of working and living and fundamentally transformed society. This process began in Britain in the 18th century and from there spread to other parts of the world.

The Industrial Revolution is a topic reviewed in schools quickly, however, techniques such as AR are favorable to complement this topic in a dynamic way.

Inventions of the Industrial Revolution in AR is an application that allows us to show through Augmented Reality different inventions developed in the Industrial Revolution, which are modeled in 3D. The most minimalist design possible was chosen to give the user a sense of easy understanding, and to feel attracted to the use of the application.

A. **Information** redirects to the scene where a panel of information of the invents of the Industrial Revolution are displayed.
B. **Gallery** shows a photo gallery of the invents of the Industrial Revolution.
C. **Help** presents the instructions of the application.

D. **AR button** shows different inventions in 3D developed in the Industrial Revolution according to the marker.

Figure 7 presents 5 inventions of the Industrial Revolution in AR. Phone, Airplane, Light Bulb, Car and the Locomotive.

Figure 7. Inventions of the industrial revolution in AR: A) Phone, B) Airplane, C) Light Bulb, D) Car, and E) Locomotive

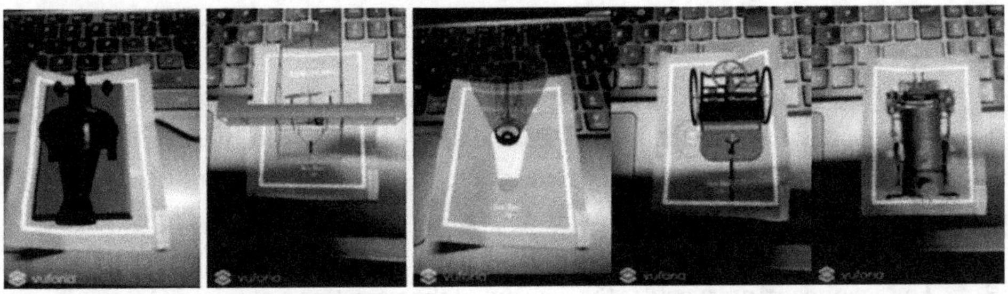

For this project, integration tests were carried out, in which the functionalities of the application were verified, such as: target recognition, display of figures, sound and the technical sheet. As well as compatibility tests, testing the performance of the software on different Android mobile devices, to verify that all the functionality offered by the application works correctly on some current smart mobile device. In addition, usability tests were carried out on 25 users from Tlaxcala, ranging from 15 to 22 years old. The following results were obtained from the "Task Orientation" Test: 85% consider that the information is presented logically and naturally, 80% take advantage of the very strengths of the user and the mobile device and 85% mention that the software makes the experience easy and quick to explore. From the "Trust and Credibility" Test, the following results were obtained: 85% consider that the software has credible content, 90% consider that the software is free of errors.

Virtual Worlds of the Tlaxcala's Diasporas

When the Viceroy don Luis de Velasco II in March 1591 signed the capitulations for the recruitment of 400 Tlaxcalan families with the purpose of founding towns in the northern New Hispanic zone threatened by the generalized extermination by the Chichimeca war, the Tlaxcalan society of the moment took Part in this process in an enthusiastic way, the friars, the council and the military structure both Spanish and Tlaxcalan. This event was so important that it initiated a process of agricultural, economic and cultural expansion and generated a source of identity and rights in the founded towns that four hundred years later still show that heritage in multiple ways. The colonization process in northern New Spain was slow and multifaceted, spanning approximately 300 years (Martínez, 2000).

This very interesting topic was devised through different virtual worlds, where the way out of the towns were presented, as well as the products that the Tlaxcalans brought to the country.

The software "Virtual worlds of the Tlaxcala's diasporas" is developed for computers, the user have the opportunity to visualize the background and the way out of the four towns: Ocotelulco, Tizatlán, Quiahuiztlán and Tepetícpac.

A. The first virtual world is the representation of the valley of the Nuestra Señora de las Nieves Convent in San Juan Totolac, where the meeting of the 400 families took place, which by royal decree of the Viceroy don Luis de Velasco (son) and with the support of the four main towns of Tlaxcala, were sent to the Great Chichimeca.
B. The second virtual world is the way out of the town San Francisco Ocotelulco, in this virtual tour the Tlaxcalan families set out on June 6, 1591, led by Captains Lucas de Montealegre and Miguel de Casas Ehcapitzactzin, the caravan was made up of 90 married couples, with approximately 250 Indians, marching towards the mines of San Andrés del Teúl, in the current state of Zacatecas.
C. The third virtual world is the way out on June 7 of the town of San Esteban Tizatlán, led by Mr. Buenaventura de Paz and Mr. Joaquín de Pedroza, who organized the caravan made up of 86 married couples from Tizatlán, with around 245 people. They went to colonize the Nuevo Reino de León and the Reino de Nueva Vizcaya, today Nuevo León, and the town of Santiago de Saltillo, today Coahuila.
D. The fourth virtual world is the way out of the town Quiahuiztlán on June 9, 1591. The Lordship of the Kings Quiahuiztlán brought together 83 families with a total of 207 neighbors and they left organized by Captain Lucas Téllez and Diego Ramírez. They were sent to the Villa de la Nueva Tlaxcala, in what is now San Luis Colotlán Jalisco.
E. The fifth virtual world is the way out of the town de Santiago Tepetícpac starting the journey with Captains Francisco Vázquez and Joaquín Paredes, 231 indigenous people. They left for the mines of San Miguel Mexquitic, in present-day San Luis Potosí.

In the caravans a concentrate of the Tlaxcalan culture was taken: a variety of seeds, fruits and plants, household items, clothing, food and traditional objects, household items that would facilitate the trip and later the settlement, as well as farm animals, such as cattle, sheep and pigs mainly. Figure 8 presents the meeting of the town Tepetícpac.

Figure 8. Way out from Tepetícpac

To carry out the tests, 40 users were summoned, 50% of them were men and 50% women. The knowledge tests were done in two stages, before and after using the software. The knowledge test prior to the use of the software consisted of 10 questions to mention the focus was on the history of the diasporas of Tlaxcala, the 4 main lordships of Tlaxcala, place where the meeting of the departure of families took place, number of families sent, date on which this event occurred, among others.

Users navigated the virtual world for one hour, after which they responded a survey of knowledge. In the prior knowledge test, the results were evident that users don't know of the history of the Diasporas of Tlaxcala, since they obtained 20% of qualification. In the second test after using the software, the user obtained 80% in the evaluation, in this case we can say that the result is satisfactory.

The Virtual World History Trip

James Watt (January 30, 1736—August 25, 1819) was a Scottish inventor, mechanical engineer, and chemist whose steam engine patented in 1769 greatly increased the efficiency and range of use of the early atmospheric steam engine introduced by Thomas Newcomen in 1712. While Watt did not invent the steam engine, his improvements on Newcomen's earlier design are widely regarded as having made the modern steam engine the driving force behind the Industrial Revolution (Bellis, ,).

The life of James Watt is interesting, as well as the technology developed by him, so representing his inventions through a virtual world was what he decided to work on, it is worth mentioning that the virtual world is set according to the time.

The virtual world History Trip presents the Technological advances developed by James Watt, the virtual world is set in a London city at the end of the 18th century, the navigation is done in the first person. It can be seen on the tour:

A. "The rocket" locomotive, one of the first steam locomotives with an 0-2-2 wheel arrangement. This to be developed by Watt implemented the improvements of the steam engine, see Figure 9.

Figure 9. The rocket modeled in 3D

B. The steam engine, an external combustion engine that transforms the thermal energy of a quantity of water into mechanical energy.
C. The copying machine, developed in 1779, used more rudimentary methods and processes than those known now, it still achieved the goal: photocopying a document. He used a special ink that passed from the original to the copy.
D. A factory, in said building the user can enter to see its interior, see and hear a machinery working, on the tables there are work utensils which the user can move.
E. Buildings such as: churches, bars, houses, schools, common areas and blacksmiths.

The tests of these project are focused on the user, observing in detail their reaction to the design and functionality of the product to be evaluated. For the usability tests in conjunction with the onlineencuesta.com platform, a survey was distributed in which the software developed in this project can be evaluated, it was distributed through a shared folder in the Onedrive cloud, said survey was distributed to 20 users where the average age is between 12-19 years old, who gave their opinion on the present work, evaluating the following characteristics:

- Initial presentation.
- Task orientation.
- Trust and credibility.
- Content and writing

Upon concluding the usability tests and analyzing the data that constitute this phase, such as the survey questions and the age range of the users, it is easy to notice the good results of this project. The interface and content of the tour are liked by users, as well as the tour itself. Returning to the hypothesis of this project, it is feasible to implement virtual tours as didactic materials, making learning more attractive and interactive with the people who use this software.

VirtUATx

Assembling a computer is useful for technicians and engineers in the computing area, so in this project it is proposed to assemble a CPU in a multi-screen environment.

VirtUATx is a multi-screen software that allows you to build different virtual environments. The screens can be placed at different angles, but the idea is that the virtual information is continuous. This environment is built with a kinect, a head tracker, as well as incorporates active and anaglyphic stereoscopy. The case study is the assembly of a computer, for which the hands and the body are used as a kind of controller. Figure 10 shows VirtUATx working.

For this project, screen continuity tests, cabinet assembly simulation, tests with the Kinect, tests of two screens with angles between them, tests of active stereoscopy and anaglyphic stereoscopy were carried out. In the same way, it was presented at different academic events, inside and outside the FCBIyT.

Virtual Worlds: The Independence of Mexico

The independence of Mexico is a basic issue in the history of Mexico, so it was decided to work on the independece, the project is based on different references, one of them is (Lee Benson, 1985). The interactive learning system using virtual worlds:

Figure 10. VirtUATx

The Independence of Mexico 1810, contains 3D scenarios where the timeline is contemplated from the beginning of the uprising in arms until the death of its first leaders, which corresponds to the year 1810 to 1811, what is considered as the first stage of the fight of the Independence of Mexico. The system consists of 6 scenarios with 3D technique and videos that provide coherence between each scene.

A. Scenario 1 presents the conspiracy in Querétaro, the main character is Mayor Ignacio Pérez, who was one of the main authors of the Independence movement. In this scenario, the user must search and reach the main door of the house of Josefa Ortiz de Domínguez. The user will learn how the city of Querétaro was constituted, specifically the environment of the house of Josefa.
B. Scenario 2 takes place in Querétaro, where the user will take the character of Ignacio Pérez, they must look for Ignacio Allende's house to deliver the alert message that the independence movement has been discovered. The user will learn the meaning of the assignment that the character had and the danger of that moment.
C. Scenario 3 is in San Miguel el Grande, the main character is Captain Juan Aldama, whose mission is to reach Dolores looking for priest Hidalgo.
D. The fourth scenario takes place in the town of Dolores, the actions taken by Miguel Hidalgo and Ignacio Allende are explained due to the independence movement was discovered, the user can tour the surroundings of Miguel Hidalgo's house and the Dolores Parish. The user plays with the character of the priest Miguel Hidalgo. See Figure 11.
E. The fifth scenario takes place in Celaya, the actions carried out by the priest Miguel Hidalgo and Ignacio Allende are explained, as well as the shout of

Figure 11. Priest Miguel Hidalgo in the Parish of Dolores

Independence. The user plays with the Character of the priest Miguel Hidalgo and must go through the stage until he reaches San Miguel el Grande.

F. The sixth scenario takes place in San Miguel el Grande, the user plays the character of Ignacio Allende and must reach the city of Guanajuato. In this scenario, the user must look for the Alhóndiga de Granaditas.

Tests were carried out with 20 users, for which five questionnaires were used, the first of which was focused on evaluating prior knowledge, after the user used the software, three usability questionnaires were applied, evaluating topics such as software presentation, navigability, content quality and credibility, with satisfactory results, finally, a knowledge test was applied again, in the initial knowledge tests against those carried out after using the project, it was possible to see that the qualification obtained with respect to the first one was higher by 30%, so it is intuited that with the use of the software the users did learn.

Video Game: Diasporas of Tlaxcala

In section 4.2 a remembrance of the diasporas of Tlaxcala has been made, the project described below takes the same theme, but now worked as a video game.

Video game: Diasporas of Tlaxcala begins with the young Cenzontle, who is interested in learning how the migration of the 400 Tlaxcalan families. Cenzontle asks his grandfather the history of Tlaxcala and begins with his grandfather's narration through a three-level game, see Figure 12.

A. **Level 1. The alliance between the Tlaxcalans and the Spanish**. During the narration it is shown how the 4 manors accepted the treaty of Viceroy Luis de

Figure 12. The grandfather narrates the history of the diasporas of Tlaxcala

Velazco. Hernán Cortés decides to explore the Mexican territory, an event that allowed him to discover the great wealth of Mexico, since he traveled a lot and encountered various dangers on his way, but when he returned to the Republic of Tlaxcala, he told him to the viceroy and the decision is made to send young Tlaxcalan couples from each manor, this produced several disagreements since they were promised benefits when migrating, but the same people were afraid of facing something unknown during the journey.

B. **Level 2. Migration From Ocotelulco and Tizatlán**. At this level, it is narrated how the migration of the people of Ocotelulco and Tizatlán arose. After this, they had discussions within the same towns, since the manors had to gather the families so that they could begin their journey towards the great Chichimeca, it was finally agreed that the families would leave Tlaxcala, on June 6, 1591 was the exit for Ocotelulco headed by Don Lucas de Montealegre. On June 7, 1591, was the exit for Tizatlán was led by Don Buenaventura de Paz and Don Joaquín de Velasco to the current territory of Jalisco.

C. **Level 3. Migration From Quiahuiztlán and Tepeticpac**. The third and last level begins with an introductory narration about of the most important events in the history of the migration of the 400 Tlaxcalan families, the exit of Quiahuiztlán and Tepeticpac, these families traveled to the north from Mexico.

The game was tested by 20 users, 80% men and 20% women, after 60 minutes of play they were given a survey with different rubrics, the results were satisfactory, obtaining 89% acceptance of the project.

REFLECTIONS

The projects presented were developed in the Faculty of Basic Sciences, Engineering and Technology, by students under my supervision, from the Computer Engineering degree and from the postgraduate courses in Computing and Electronics. During the development it was possible to perceive that either Virtual Reality, Augmented Reality or video games, are powerful techniques that allow excellent results to be obtained, appealing to all types of users, especially for children and young people, who are already immersed in technology. The projects were each tested with modular, integration and usability tests, in order to have quality projects. Many of them were also tested to verify that new knowledge had been obtained. In general, from the tests we can see that the projects have a positive approval. In general, users stated that it is in their interest to use the different projects shown in this chapter. It can be said that the work carried out at the FCBIyT is creative, innovative and always seeks to support society. The quality of the projects speaks of the quality of the graduates at the Autonomous University of Tlaxcala

CONCLUSION

We can say that AR is currently very common among young people, so developing applications, whether in medicine or education, has been feasible, since from their cell phones young people can learn about these issues and at the same time be evaluated to guarantee a learning with this technology. On the other hand, RV has been achieving an interesting evolution, given that computer equipment now better supports heavy graphics, which before only large and sophisticated equipment could support. So VR is becoming more accessible for students and therefore, it can be very useful for teaching medical and educational aspects. Finally, video games that involve a lot of work, from design, story, interaction, multidisciplinary team work, are also beginning to become accessible to develop by a group of students with programming and design knowledge. These video games are possible and accessible for students who want to complement their knowledge of various medical situations, as well as education.

RECOGNITION TO THESIS STUDENTS

These projects presented were developed at the Faculty of Basic Sciences, Engineering and Technology, under the advice of Dr. Marva Mora Lumbreras, with the following work team:

1. Maribel Cuamatzi Muñoz (Arachnophobia)
2. Britany Alejandra Moreno Cruz y Raul Alejandro Hernández Pérez (TEAPRENDE)
3. Carlos Alberto Maldonado Flores (SARA)
4. Esther Ortega Mejía (AVHI: Non-Hodkin Lymphoma)
5. Elizabeth Zarate Sanchez (Byoky)
6. Juan Pedro Morales Pluma y Orlando Vazquez Muñoz (LenSeñasMx)
7. Ricardo Rafael Ramos Ordoñez (Inventions of the Industrial Revolution in AR)
8. Yesenia Velazquez Corona (Virtual worlds of the Tlaxcala's diasporas)
9. Jair Méndez López y José Omar Méndez García (Virtual world History Trip)
10. Alvaro Jair Martínez Varela (VirtUATx)
11. Carlos Alberto Meza Alvarado (The interactive learning system using virtual worlds: The Independence of Mexico 1810)
12. Diasporas of Tlaxcala

REFERENCES

ASD. (2023). *Autism Spectrum Disorder, Signs and Symptoms of Autism Spectrum Disorder*. CDC. https://www.cdc.gov/ncbddd/autism/signs.html

Bellis, M. (2023). *Biography of James Watt, Inventor of the Modern Steam Engine*. ThoughtCo. https://www.thoughtco.com/james-watt-inventor-of-the-modern-steam-engine-1992685

Bourdon, K. H., Boyd, J. H., Rae, D. S., Burns, B. J., Thompson, J. W., & Locke, B. Z. (1988). Gender differences in phobias: Results of a ECA community survey. *Journal of Anxiety Disorders*, *2*(3), 227–241. doi:10.1016/0887-6185(88)90004-7

Britannica. (2023). *Industrial Revolution*. Britannica. https://www.britannica.com/event/Industrial-Revolution.

Faurot, K., Dellinger, D., Eatough, A., & Parkhurst, S. (1992). *The identity of Mexican sign as a language*. SIL International.

Front labeling of food and beverages. (2021, October 5). gob.mx. https://www.gob.mx/promosalud/acciones-y-programas/etiquetado-de-alimentos

Martínez, S. T. (2000). *The Tlaxcalan Expansion to the New Spanish Colonial North: The Stages of Agricultural Expansion*. Colegio de Postgraduados. https://www.researchgate.net/publication/269410739_la_Diaspora_Tlaxcalteca

Mayo Clinic. (n.d.), *Dissociative disorders*. Dissociative Disorders. https://www.mayoclinic.org/diseases-conditions/dissociative-disorders/symptoms-causes/syc-20355215

NCI. (2022). *National Cancer Institute, Adult Non-Hodgkin Lymphoma Treatment (PDQ®)–Patient Version*. NCI. https://www.cancer.gov/types/lymphoma/patient/adult-nhl-treatment-pdq,

Lee Benson, N. (1985). *Mexico and the Spanish Cortes, 1810-1822. Eight Essays, Aus tin, University of Texas Press, 1966, que fue traducido al español: México y las Cortes españolas 1810-1822*. Ocho ensayos, México, Instituto de Investigaciones Legislativas, Cámara de Dipu tados, 1985.

NLM. (2022). *Capgras Syndrome*. National Library of Medicine, National Center for Biotechnology Information. https://pubmed.ncbi.nlm.nih.gov/34033319/

Rothbaum, B. H., Hodges, L. F., Ready, D., Graap, K., & Alarcon, R. D. (2001). Virtual reality exposure therapy for Vietnam veterans with posttraumatic stress disorder. *The Journal of Clinical Psychiatry*, *62*(8), 617–622. doi:10.4088/JCP.v62n0808 PMID:11561934

Smith-Stark, T. C. (1986). *The Mexican manual language* [Unpublished paper]. Colegio de México.

Strategic National Programs. (n.d.). *Programas Nachionales Estraegicous*. SNP. https://conacyt.mx/pronaces/

Chapter 10
A Theoretical-Practical Case Study on the Graded Multidisciplinary Model:
Training of High School Students Through STEAM Education

Mauricio Flores-Nicolás
https://orcid.org/0000-0002-7172-8272
Universidad Autónoma del Estado de México, Mexico

Magally Martínez-Reyes
https://orcid.org/0000-0002-2643-6748
Universidad Autónoma del Estado de México, Mexico

ABSTRACT

A model of educational processes called multidisciplinary gradual is presented, whose objective is to guide the teacher in the teaching-learning of STEAM educational activities using as a use case the topic of basic logic gates, which associates mathematical, algebraic, electronic, and logical knowledge, presenting the theory and strengthening it in practice. The topic of logic gates is not directly related to the middle-higher level educational plan; however, it is possible to apply and deepen it by relating the basic mathematical operations. The work describes the three stages that make up the model, achieving that students approach a problem or topic from different contexts, each discipline contributes its own knowledge and qualities, but not necessarily integrated in the common understanding of the problem, so that they develop the foundations to solve interdisciplinary and/or transdisciplinary problems. At the end, the results obtained by using this model with data from a control group and an intervention group are demonstrated with statistical analysis.

DOI: 10.4018/978-1-6684-8639-9.ch010

INTRODUCTION

Nowadays, education is in constant evolution due to the integration of new technologies, therefore there is a need to develop 21st century skills such as creativity, computational thinking, argumentation, among others, in addition to strengthening students' knowledge. The STEAM approach, which combines Science, Technology, Engineering, Art, and Mathematics, has become a more effective way of teaching to achieve this goal. The inclusion of technology in education is fundamental to preparing students for a digital world; however, it is also important to use it efficiently, effectively, and responsibly. In addition, the collection of statistical data analysis allows to evaluate the impact of these educational practices and to continuously improve STEAM teaching. In this context, it is essential to analyze how to integrate technology in education, to develop skills and to evaluate the impact of these practices through statistical data. The Graded Multidisciplinary Model (GMM) promotes the use of different pedagogical methods, educational strategies, and technologies to guide teachers in the design and instruction of classes with a STEAM approach.

BACKGROUND

STEAM education is a paradigm with the goal of developing skills that depends on three approaches: 1) Interdisciplinary, this involves the collaboration of different disciplines, each contributing content from their own point of view and specific skills to address a complex topic or problem (Conde et al., 2021; Greca, Ortiz-Revilla & Arriassecq, 2021); 2) Transdisciplinary, focuses on transcending disciplinary barriers and exploring the topic from a global perspective by incorporating both expertise and knowledge from various disciplines (Li et al., 2020; Peterson & Hipple, 2020; Huffman, Thomas & Basham, 2020); and 3) Multidisciplinary, which involves combining different disciplines, knowledge and qualities but independently to address the complex issue or problem (Flores-Nicolás & Martínez, 2022; McLure, Tang & Williams, 2022). STEAM aims to foster 21st century skills (creativity, critical thinking, logical-mathematical thinking, problem solving, among others) to prepare students to face current and future challenges to complex and unknown problems (Conde et al., 2021; Krüger & Chiappe, 2021).

The importance of STEAM education lies in facilitating the understanding and application of knowledge from traditionally separate disciplines to develop the skills needed for the jobs of the future in an increasingly technological and changing world (Li et al., 2022; Wang et al. 2022). In addition, it seeks to improve diversity and inclusion with the participation of women and minorities in these fields to reduce the gender and ethnic diversity gap (Nasri et al., 2021). This is due in part to the inclusion of the arts, allowing students to gain a different perspective, not only to maximize professional or academic opportunities, but also in personal and social development (Lorenzo-Lledó, 2023). Broadly speaking, STEAM education helps to make the development of STEM subjects more attractive, achieving a personal improvement to expand professional opportunities.

On the other hand, there are several challenges that content instructors and educational institutions face when designing, developing, and implementing activities in STEAM education, below are five of the most common challenges. 1) Lack of resources, teaching-learning with STEAM educational strategy requires access to specialized equipment and/or technology (Jackson et al., 2020; Flores-Nicolás & Martínez, 2022), with a complexity-knowledge relationship. In other words, for example, if a teaching kit comes perfectly documented for the student and the teacher, and delimited with the different activities that can be performed step by step, its cost will be higher, because the only thing that must be done is to read such material to achieve the objective, significantly decreasing the complexity; on the other hand, if a kit is delivered with different electronic components (breadboard, resistors, LEDs, cables, among others) whose documentation depends only on the use and operation of such components without specifying which activity or practice can be performed, such material will have a lower cost, however, the complexity will increase considerably (Catlin, Kandlhofer & Holmquist, 2018). 2) Limited teaching training/experience, giving continuity to the previous challenge, most teachers do not have the necessary training and/or experience to effectively teach activities in STEAM education for their curriculum (Avdeenko, Sabirova & Konyushenko, 2021; Margot & Kettler, 2019), giving two possible scenarios: the first, if the academic institution is located in a low-resource area, ideally the most economical material should be procured; however, this requires teacher training to be increased to effectively and efficiently take advantage of such material; and second, if both the institution and the students have economic resources, teacher training will be limited to following the manuals to achieve the objectives; this second case in particular is of some concern because apart from having expensive material, the skills and knowledge that STEAM promotes are not used (Flores-Nicolás & Martínez, 2022). 3) Difficulty in evaluation, evaluating activities in STEAM education can be difficult, since new indicators are required that take advantage of the skills of the 21st century, apart from traditional methods, such as the application of exams (Margot & Kettler, 2019; Urgiles et al., 2021). 4) Diversity and inclusion, STEAM education also encourages all students to have equal opportunities to learn and develop their skills, regardless of their background, gender, race, sexual orientation, or any other personal characteristics, diversity in the classroom allows for a greater variety of perspectives, which enhances 21st century skills and their application in learning; in addition, the inclusion of students from different backgrounds fosters tolerance and respect towards cultural diversity and promote social justice (Areljung & Günther-Hanssen, 2022; Lorenzo-Lledó, 2023; Alba, 2019). 5) Finally, in STEAM education, failure is an important part of the learning process, because when working on complex and challenging projects, students themselves face failures and mistakes, instead of seeing failure as something negative, it is important to foster a positive and learning attitude towards it, the purpose of promoting this is that students see mistakes as opportunities to learn and grow, to develop their resilience, their ability to solve problems and their creativity; In addition, it can decrease the fear of failure and foster a collaborative and supportive work environment in which all students support each other during their learning, feeling more comfortable taking risks and facing new challenges (Jackson et al., 2020; Flores-Nicolás & Martínez, 2022).

To overcome these challenges, schools may need to invest in resources, provide teachers with training and professional development opportunities, work with community partners to increase resources and expertise. It is important for schools not to be closed to change, such as new methods of assessment and the use of technology for teaching-learning, and above all to ensure that the curriculum is accessible and diverse for students.

Some findings on how to acquire such STEAM education competencies, 21st century skills and overcome some challenges, can be found in the work of Conde et al. (2021), where they mention that through active methodologies such as the Systematic Mapping Process (SMP) is an interesting way to involve students in addressing subjects such as robotics and mechatronics. An example before applying STEAM education, Ohnishi et al. (2017), identified and applied specific kits according to the educational level (primary and secondary), with a theme of "cleaning robots", this encouraged children to think proactively, given that cleaning is familiar to them; however, given the nature of programming, there was not enough time available in the classroom, so they suggest adopting interesting educational material for effective teaching-learning of programming and robotics. In another manuscript Chang, Debra & Chang (2019), developed and implemented XR (augmented, virtual, mixed, or extended reality) based modules and manuals to establish an online platform to assist teachers in sharing knowledge and experiences, and provide feedback on XR technology in organizing workshops in teaching-learning STEAM education. On the other hand, Liston et al. (2022), present a collaborative work using an environmental monitoring system based on the Internet of Things (IoT) and data science, using the STEAM education approach, collected qualitative data through questionnaires and interviews to understand how participants interpreted their experiences and what meaning they attributed from them.

Meanwhile, Clark et al. (2015), related the theory and design of digital games to support the development of scientific competence in children applying two theoretical frameworks Knowledge in Pieces (KiP) and Science as Practice (SaP), arguing the importance of disciplinary integration through games for science learning. Something similar happened with Miller (2018), he analyzed some math apps on iPads on number sense learning in game-based learning environments within a preschool, he observed how interactive technology promoted collaboration and student engagement although he did not obtain significant academic results, because children preferred more creative and entertaining apps, the difficulty or complexity of the use of the app and content also affected their engagement. On the other hand, Arufe et al. (2022), study how the use of gamification in a multimodal learning environment influences the academic performance of university students, in a use case with two groups (control and intervention), explaining that the intervention group obtained better results, both in grades and tasks, while the control group did not obtain changes when receiving traditional teaching. Likewise, Jensen & Hanghøj (2020), used Minecraft as a tool and learning environment, which served as a support for teaching mathematics by applying the Minecraft coordinate system, the result of interviews with students showed that they experienced new ways to participate in mathematics, the author concludes with design principles for the future use of computer games for mathematics education.

These works are based on the design and instruction of didactic materials to develop activities and practices based on STEAM education. The research is classified into three types: 1) aimed at teachers, with proposals ranging from the creation of blended learning environments (Yin & Yuan, 2021), didactic-technological integrated instructional materials (Tekerek & Tekerek, 2018), design of conceptual models with the educational strategy of problem-based learning (PBL) and transdisciplinary approach (Quigley, Herro & Jamil, 2017), design and application of project-based learning (Juškevičienė, Dagienė & Dolgopolovas, 2021), and application of the CACAO model for the transformation of teaching culture and practices in STEM departments (Viskupic, Earl & Shadle, 2022); 2) aimed at students, such as the work of Chung et al. (2022), which aims to integrate imagination and STEAM education through a special e-science course using mobile devices at the secondary level (Chung et al., 2022); and 3) aimed at both interdisciplinary approach and project-based learning to effectively discuss and link disciplines (MacDonald et al., 2020).

The proposal is to use the GMM that was originally designed to create educational software; however, by making modifications in the processes, it is possible to create a combined strategy (techno-pedagogical) aimed at three or more disciplines of STEAM education. Teaching-learning with activities of this nature is a crucial moment in the formation of students since it is when they begin to choose their academic and professional future. Among the skills expected from students before entering the University are critical, creative, and analytical thinking to solve problems and the ability to innovate, by making the solutions they propose creative to complex problems. The suggested topic is about logic gates because it can be presented at different levels of understanding, so it will help students to focus and continue to deepen the topic in different disciplines. In the following section, the procedure that allows designing and implementing educational activities with STEAM nature is described.

METHODS AND MATERIALS

The research was conducted in two parts (Construction of educational activities in STEAM, and evaluation), which consist of the following:

Planning: The GMM brings together knowledge, techniques and strategies coming from different disciplines to analyze and unify them, managing to identify relevant aspects about STEAM education to design and build educational activities of this nature, and create opportunities for improvement. Unifying the educational perspective with the technological one requires a triple process, described below:

1. Elaboration of a Hypothetical Learning Trajectory (HLT). It refers to a possible training path that a student or group of students may follow to achieve certain educational objectives. The route is hypothetical because there is no single way to achieve these objectives; they vary according to the needs, abilities, and preferences of the students, as well as the resources and tools available (Ivars, Fernández & Llinares, 2020; Flores-Nicolás & Martínez, 2022). The elaboration of the HLT requires documentary research of the topics related to logic gates, most of the content will be addressed with the books Electricity and Electronics by Rela (2010) and The Boolean Algebra and its applications by Whitesitt (2012), (see Figure 1a).
2. Educational Systems Engineering (ESE). This process analyzes important aspects of the technology to be used (hardware and software), such as: what interaction will the technology have with the user? In what activities will it be used? Questions such as: what the minimum training is required for handling the technology (teacher), costs (teacher or student), work area (institution), among others, are answered. The main point of the procedure is to effectively place the technology in STEAM activities; it should be clarified that it is not expected to cover the entire activity, only a part of it, so that it is possible to offer different alternatives (Schuelka & Thyrring, 2022; Flores-Nicolás & Martínez, 2022). The analysis through the ESE, will be performed with two technologies: the first, Minecraft Education (software) (see Figure 1b), whose premise is simple, such as extracting materials to use them in the construction of multiple things, it is also multiplatform, being accessible from mobile devices,

to video game consoles; and the second, an electronic device (hardware) (see Figure 1c), this is a large-scale prototype of the connections of integrated circuits (IC) 74LS04, 74LS08 and 74LS32.

3. Instructional Design (ID). Finally, a systematic process is used to plan, develop, implement, and evaluate teaching-learning programs. The two previous elements (HLT and ESE) are combined to create a structured and research-based program, in this way, instructional designers ensure that the educational programs are adaptable to the needs of students and learning objectives, to obtain measurable and evaluable results, this is important, because authors such as Juškevičienė, Dagienė & Dolgopolovas (2021), mention that statistical evaluations are lacking in this type of research (Avalos & Castellanos, 2022; Flores-Nicolás & Martínez, 2022). All the information obtained will be concentrated in the planning format (see Figure 1d).

Figure 1. Minimum components for teaching-learning the topic of logic gates with a focus on STEAM education
Source: Own elaboration.

a) Boolean Algebra and its Applications by Whitesitt (2012), and Electricity and Electronics by Rela (2010).

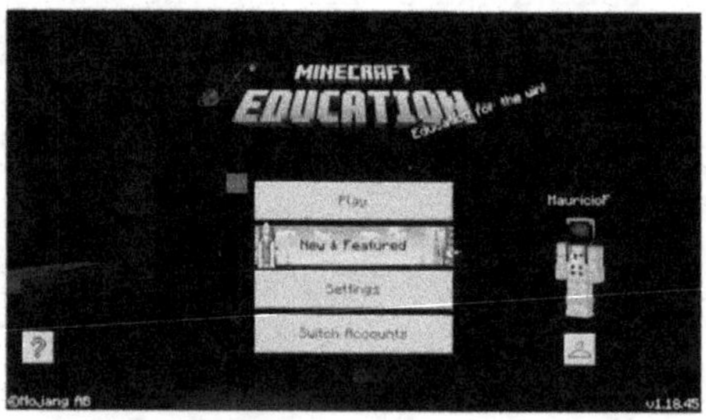

b) Minecraft Education.

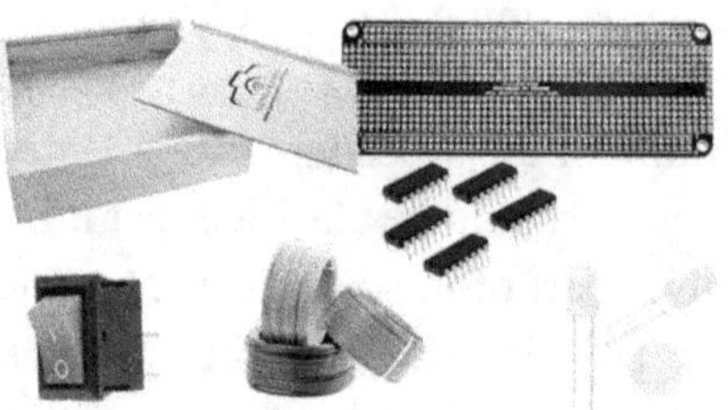

c) Materials for the construction of the electronic device.

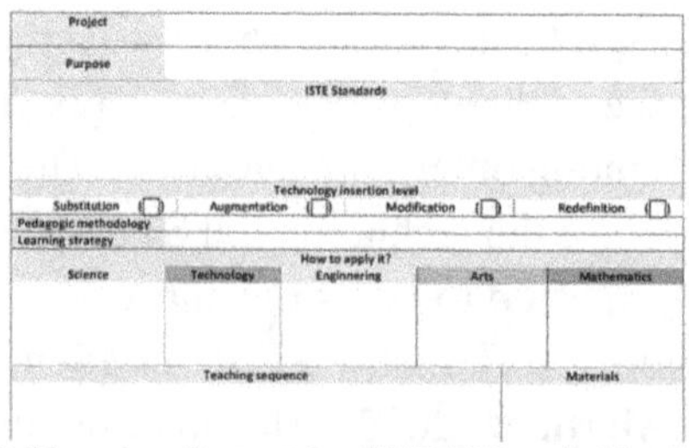

d) Planning format for STEAM activities by Flores-Nicolás & Martínez (2022).

Evaluation: This research uses the pre-test and post-test techniques to measure the effectiveness of didactic instruction during class. This technique involves measuring a variable before (pre-test) and after (post-test) the instruction. By comparing results, it is possible to determine if there was any significant change in a variable of interest. In addition, it is proposed to use a group of participants who do not receive the educational intervention being tested. The purpose of including a control group is to allow comparison between the control group and the experimental group (the one with the educational intervention) to determine whether the intervention has a significant effect. Data collection is a critical and essential step in providing feedback and improving the GMM (Martinez, 2006).

DEVELOPMENT

Based on the components of the previous section, the development of the three processes that make up the GMM will be described to obtain the ID in STEAM activities whose main theme is logic gates.

HLT for the Teaching-Learning of Logic Gates

For the first stage of the GMM, a topic known by the students of this educational level will be presented, but in another context, starting with simple mathematical operations such as addition and multiplication until reaching their respective concept and representation in Boolean algebra of the same, to achieve this objective, the two technological devices will be used as intermediaries.

Figure 2 is a graphical representation of what is known as a HLT with gradual and multidisciplinary characteristics (Flores-Nicolás & Martínez, 2022), because in the distribution of the outer nodes is the content whose understanding requires prior knowledge of the students; that is, with this HLT it is possible to introduce students on the subject in any of the STEAM disciplines. The distribution of the content, as well as their respective sources, is described below: From Whitesitt's book (2012), chapter 2 was used for the Mathematics and Boolean Algebra node, and part of chapter 3 was used for the Sciences node (Logic and its symbology); from Rela's book (2010), the Engineering node (Digital Systems) was designed and the electronic device was built for the Technology node (Hardware); For the Arts node, a combination of the two books was made; finally, for Technology, the respective analysis will be described in the following subsection, with the purpose of defining when it would be appropriate to use it with the students, since it will serve more as a support tool and not as the main attraction of the subject.

Figure 2. Design of the HLT for high school level students on the subject of logic gates
Source: Own elaboration based on the design of Flores-Nicolás & Martínez (2022) and the content of Whitesitt (2012) and Rela (2010).

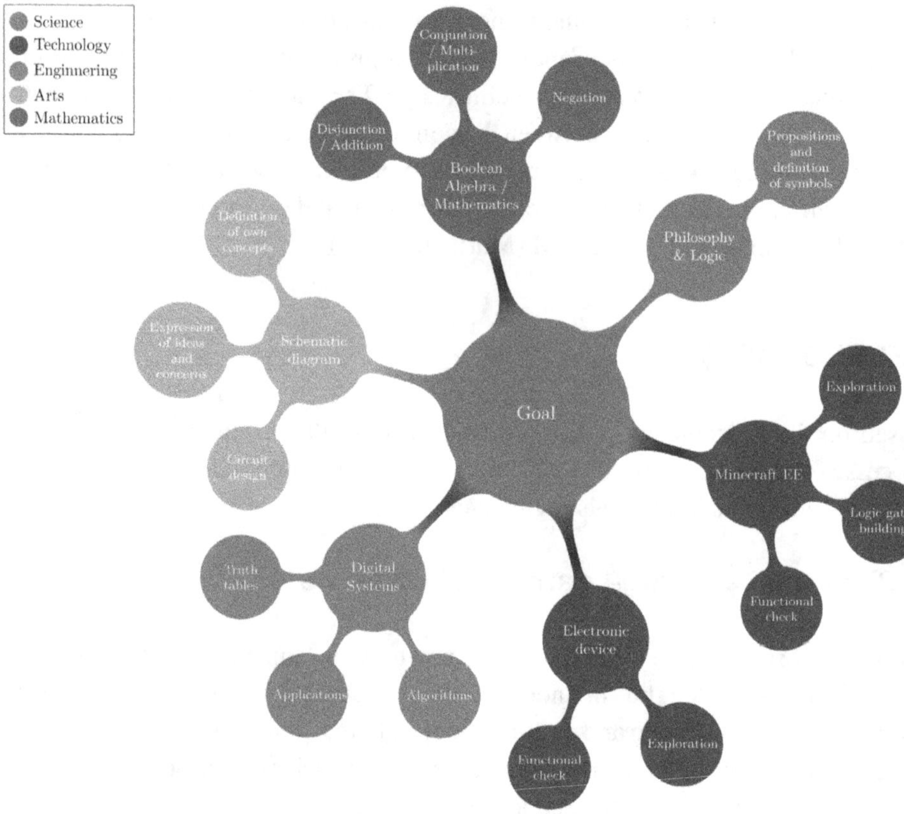

Technology Analysis

Due to the requirements represented in the HLT (see figure 2) in using two types of technology (Hardware and Software), this subsection will be divided into two to adequately address them, described below:

Minecraft Education. It is a popular open world video game developed by Mojan Studios, since its appearance in 2011, has been improving to attract millions of players around the world. The motivation of the game is that players can explore and build with the materials at hand, for that requires activities ranging from logging wood, extract minerals to raise animals for food. Minecraft is compatible with a variety of devices and platforms, making it accessible to its wide audience, among these compatible technologies are: PC, in versions available for Windows, macOS and Linux; Video game consoles, such as Xbox 360, One, S|X Series, PlayStation 3, 4 and 5, Nintendo Wii U and Switch; Mobile devices, on Android, iOS and Windows Phone systems (although it is already discontinued it is possible to perform different educational activities); Virtual Reality, it is also compatible with some headsets such as Oculus Rift and Windows Mixed Reality; and finally, Minecraft Education, available for Windows, macOS, iOS, Android and Chromebook. The latter, offers a trial of 15 uses or logins, during that period, the teacher and students can use all the functions of the video game, if for some reason that limit is exceeded, it can be purchased for $5.04 USD per year, with discounts to educational institutions (Mojan, 2023).

As the educational version of Minecraft is the one to be worked on, Table 1 shows the minimum version of the OS compatible per device, while Table 2 details the minimum technical specifications in its PC version, so that the software can run moderately well. Based on the minimum technical and electronic characteristics required for optimal execution of the video game, it was decided to use an Intel NUC workstation with Intel Celeron N3050 processor, 256 GB storage and 8GB DDR3L RAM with a Windows 11 Pro OS. In addition, the wireless sensor was

Table 1. Minimum OS requirements

Device	Minimum OS supported
Chromebook	Chrome 83
macOS	Catalina 10.15
PC	Windows 8
Android	Oreo 8.0
iPhone / iPad	iOS 11 / iPadOS 11

Source: Justin (2023).

Table 2. Minimum hardware requirements for Chromebook, macOS and Windows

Hardware	Required specification
CPU	Intel Core i3-3210 a 3.2GHz
RAM	2 GB
GPU	Integrated: Intel HD Graphics 4000 or AMD Radeon R5 series with OpenGL 4.4
Discrete	Nvidia GeForce 400 Series or AMD Radeon HD 7000 series with OpenGL 4.4
Storage	1 GB for the game core, maps, save files and others.

Source: Justin (2023).

used to replace the keyboard with the Xbox 360 controller on the PC, this decision was before the end of the analysis, since it was concluded that the students would not be sufficiently motivated if they played with the keyboard. Teacher training can be found for free on Microsoft Learn (2023), with courses ranging from basic Minecraft basics to using block programming.

Basic logic gates trainer. Unlike the electronic device proposed by Flores-Nicolás & Martínez (2022), where the student oversees connecting the electronic components to make it work, here the role of technology is as a support to check the operation and perform operations in the different representations (electronic, algebraic, mathematical and logic). The original electronic components and complementary circuits that will be used to build the trainer are shown in Table 3. The price compared to the original proposal is higher because this should be simplified so that students understand the subject, instead of involving them in the construction of their own circuits (although understanding the subject, it would be possible), with this it is highlighted that this type of circuits falls into the classification of educational robotics described by Catlin, Kandlhofer & Holmquist (2018), and agrees with its cost-complexity ratio. It is recommended that, if the teacher does not have the necessary knowledge to build this type of circuits, he/she should turn to an expert or advisor to guide him/her during the construction of this, to solve some technical problem, at the same time that he/she is trained in its use. Figure 3 shows the first phase of construction of the basic logic gates trainer, this consists of designing a chassis with the appropriate measures so that the electronic components fit and

Table 3. Cost of materials for the construction of the electronic device

Electronic component	Quantity	Price per unit (USD)
Integrated circuit (IC) 74LS04 inverter	1	$ 0.960
IC 74LS08 AND gate	1	$ 0.840
IC 74LS32 OR gate	1	$ 0.800
Generic 5-volt (V) to 1 ampere (A) charger	1	$ 2.333
Resistance of 560 ohms (Ω)	5	$ 0.013
Light emitting diode (LED)	5	$ 0.054
Banana jack (male)	20	$ 0.400
Banana jack (female)	50	$ 0.340
Switch	3	$ 0.670
14-gauge wire (meter)	3	$ 0.540
MDF chassis	1	$ 4.300
Total:		$ 40.628

Source: Own elaboration.

Figure 3. Electronic device for teaching-learning of logic gates
Source: Own elaboration.

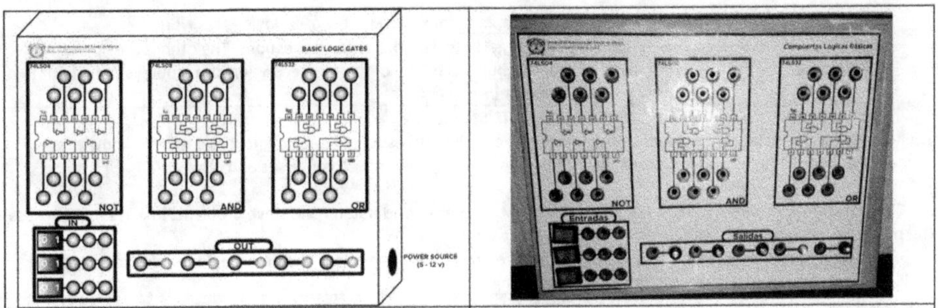

a) Mock-up of the electronic device. | b) Basic logic gates trainer completed.

have enough space if it requires future maintenance. The mock-up is divided into six work areas: the first, consists of the possible connections to be made by the NOT gate (74LS04 IC); the second, contains the AND gate connections (74LS08 IC); the third, contains the OR gate connections (74LS32); the fourth, consists of providing three input signals that will change from zero (0) to one (1) through the switch; the fifth contains the LEDs that serve as actuators and will visually notify the student when the output of the gate is zero (LED off) or one (LED on); and the sixth, a connection at the bottom right to supply power through the generic charger (see Figure 3a). Once the connections on the mock-up have been reviewed and assessed, we proceed to make the physical connection with the real materials (see Figure 3b), thus finishing with the construction of the basic logic gate trainer. It is important that teachers who want to use the technology and its does not exist, get involved so that at the same time they are trained in tis use.

STEAM Activity Planning

The last GMM process aims to create and develop effective learning experiences from the previous two stages by systematically analyzing and planning the available resources. Figure 4 shows the basic components of a ID, such as: the name of the activity/project, the name should be appealing to the learners; the pedagogical method, in this case constructionism; the educational strategies, conceptual change, PBL, simulation and gamification were chosen: the materials, these include the indispensable ones such as blackboard, markers, slides, questionaries and technological tools

(Minecraft Education and the basic logic gates trainer); the educational resources to be planned, based on the design of the HLT of stage 1 and structured through a

Figure 4. Educational planning for the teaching-learning of logic gates
Source: Own elaboration based on the format of Flores-Nicolás & Martínez (2022).

Project	Simulation vs. Reality: Guessing game about how basic logic gates work.
Purpose	Have students build logic gates to ask true or false questions. The other participants can answer using switches or buttons of any of the two technological resources to find out if the answer was correct or incorrect.

ISTE Standards
1.3 Knowledge Constructor. Students critically evaluate a variety of resources using digital tools to build knowledge, produce creative artifacts, and develop meaningful learning experiences for themselves and others. **1.5 Computational Thinker.** Students develop and use strategies to understand and solve problems in ways that harness the power of technological methods to develop and test solutions. - 1.5.c. Students break problems into component parts, extract key information, and develop descriptive models to understand complex systems or facilitate problem solving.

Technology Insertion Level			
Substitution (X)	Argumentation (X)	Modification (X)	Redefinition (X)

Pedagogical methodology	Constructionism
Learning strategy	Conceptual change; Problem-based learning; Simulation; Gamification.

How does it apply?				
Science	**Technology**	**Engineering**	**Arts**	**Math**
Use and development of logical-mathematical and logical-computational thought, basic foundations of circuit theory.	Favor decision-making with the choice of the technological tool that allows you to solve the problem.	Know the fundamentals of manufacturing electronic components, such as design, simulation, manufacturing and testing.	Designers use their creativity to create complex circuits that solve specific problems, plus circuit design itself is a visually appealing process.	Theoretical verification of the operation exposed in the technological tools, used to describe the relationship between inputs and outputs of a digital circuit.

Didactic sequence	**Materials**
Opening (20 minutes) 1. Presentation of the teacher to the class. 2. Ask students if they know Minecraft. 3. Ask students about logic gates. 4. Carry out pre-test. **Development (50 minutes)** 1. The teacher presents the objectives of the activity. 2. It briefly explains about Minecraft and the basic logic gates trainer. 3. Definition and importance of logic gates. 4. Definition and theoretical-practical operation of the NOT gate in its representation of Boolean algebra ($\neg$), mathematics ($\bar{A}$) and electronics (CI 74LS04). 5. Solve doubts and errors. 6. Definition and theoretical-practical operation of the AND gate in its representation of Boolean algebra ($\wedge$), mathematics (*) and electronics (CI 74LS08). 7. Solve doubts and errors. 8. Definition and theoretical-practical operation of the OR gate in its representation of Boolean algebra ($\vee$), mathematics (+) and electronics (CI 74LS32). 9. Solve doubts and errors. 10. Carry out the post-test. **Closing (15 minutes)** 1. Make a synthesis of what was discussed in class, highlighting the importance of logic gates for everyday life. 2. Ask about their experience using Minecraft and the Basic Logic Gates Trainer in the classroom. 3. Farewell.	• Diagnostic evaluation (pre-test). • Final evaluation (post-test). • PC with Xbox 360 controller set up and Minecraft EE (Basic Logic Gate Map). • Basic logic gates trainer (extension plug, power supply, male banana plug to male banana plug cables). • Slideshow.

Assessment	Data collection through pre-test and post-test, with algebraic and mathematical operations exercises using technological tools for verification.

didactic sequence; and the evaluation of the teaching-learning process (pre-test and post-test). However, being activities in STEAM education, some additional parameters are included, such as: the ISTE standards (2023), which provide the competencies for learning, teaching and leading in the digital age, providing comprehensive routes for the effective use of technology; the SAMR model, the first two components (Substitution and Argumentation) seek to improve learning activities with the intentional use of technology, while the remaining two components (Modification and Redefinition) are intended to transform the activities that teachers already performed traditionally; Finally, as the activity is applied in STEAM, boxes are created where it should be mentioned how the activity is influencing the five STEAM disciplines (at least three to belong in this classification), with the purpose that both teachers and students identify the importance of the topic in more than one discipline and how it can influence if it is approached with different approaches.

Design of Measuring Instruments

The first measurement instrument (pre-test) consists of a diagnostic test where eight questions are formulated, four of them are multiple choice, questioning about video games for teaching, gates, and logical propositions, the remaining four are open questions consisting of basic operations using addition and multiplication with binary numbers.

The second measurement instrument (post-test) is divided into two parts: the first, open questions consisting of identifying and operating the logic gates in their different representations (algebra, mathematics, and electronics); and the second, multiple-choice questions on satisfaction with the activity. It is emphasized that only the post-test was used with the control group, because this allows determining whether the changes observed in the intervention group are due to the educational performance or to other factors.

RESULTS

The following is a synthesis of the data obtained, distributed in four subsections: the first consists of showing the results with the most relevant questions (five from the first part and three from the second) of the post-test carried out by the control group; the second consists of data obtained from the intervention group through six questions from the pre-test and the same eight questions from the post-test; the third consists of an analysis of the two post-test questionnaires of both the control group and the intervention group; and the fourth consists of some observations and recommendations that arose during the experience in the intervention group.

Control Group Data

The questionnaire was completed by 85 students, who provided a wealth of valuable and useful data for the study at hand. Data were collected on multiple variables, including age, gender, grade, and previous experience with the topic of interest. In addition, students' opinions and attitudes regarding different aspects related to the topic, the way of teaching and the tools used were evaluated, which provided a more complete picture of their knowledge and perceptions. Figure 5 shows a graph with the responses obtained, classifying the information in two: first, question 1, 2, 6, 7 and 8 by having binary options (true, false of zero, one) there is the possibility that respondents have guessed, especially if they are not sure of the correct answer, this conclusion was obtained because there is not much difference in percentages between correct and incorrect results, to rule out or confirm the above, there is another classification of results; the second, the one composed of open questions 3, 4, and 5, where it is observed that the control group does not have the fundamentals of basic logic gates nor was the intervention offered or did they have adequate prior training, within the STEAM educational approach, there is also the possibility that the control group did not have a level of motivation or interest in the subject.

Figure 5. Percentages of correct and incorrect answers about logic gates and their operations in the control group
Source: Own elaboration.

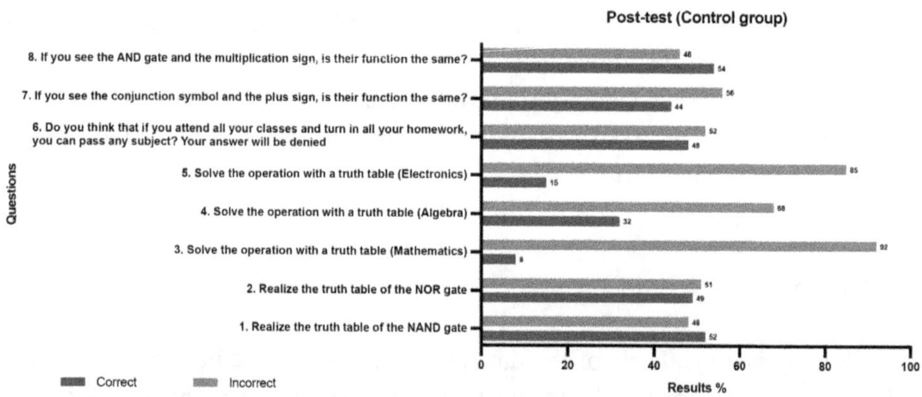

Pre and Post-Test Data of the Intervention Group

A study was conducted with a group of 26 high school students, where both the pre-test and post-test questionnaires were applied, with the objective of evaluating the effectiveness of the tutoring program elaborated with the stages of the GMM. The capture of pre-test and post-test information is essential in any research to determine the effectiveness of the intervention. The objective of the pre-test questionnaire was to measure students' prior knowledge and skills that indirectly relate to basic logic gates across different disciplines. The data obtained from the pre-test show that the group does not know the topic of logic gates, but they do have the mathematical and logical context. Figure 6 summarizes the information from the students' diagnostic test, described below:

Figure 6. Percentages of students' logical-mathematical perception
Source: Own elaboration.

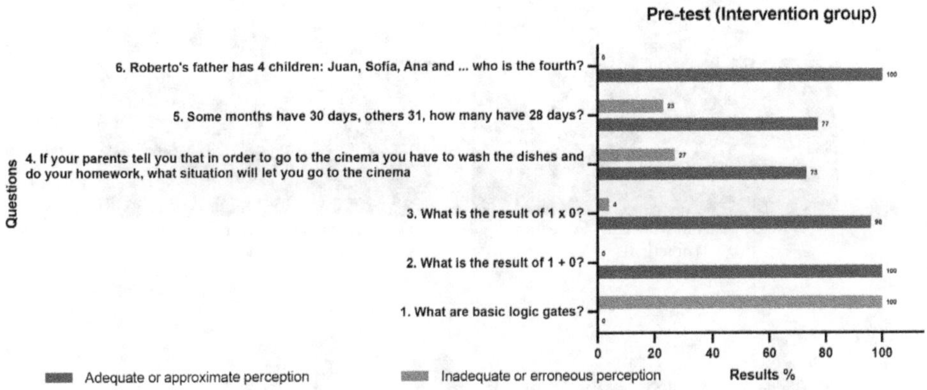

1. 100% do not know the concept of basic logic gates, their operation and symbology.
2. 100% know how to perform an operation using the plus symbol (+).
3. 96% knew how to perform the operation using the multiplication symbol (x), while 4% had an erroneous perception.
4. 73% have an adequate perception of the use of operators (and, or) in logical propositions, while the remaining 27% had an inadequate perception.
5. 77% have an approximate perception about the principles of correct thinking and reasoning through logic, while the remaining 23% had an erroneous perception of the question.
6. 100% were able to answer the question adequately.

Next, Figure 7 shows some evidence captured through photographs of the exhibit on logic gates, according to the educational planning in Figure 4.

The objective of performing the educational intervention prior to the post-test is to evaluate the effect of the intervention through the study group. On the other

Figure 7. Intervention group (Educational experience orchestrated through the GMM)
Source: own elaboration

a) Solve the pre-test.

b) Logic gates (theory).

c) Logic gates in Minecraft Education (practice).

d) Logic gates in the electronic trainer (practice).

e) Traditional feedback.

f) Feedback with Minecraft Education.

g) Feedback with the electronic trainer.

h) Solve the post-test.

hand, the objective of the second questionnaire is to measure the learning outcomes obtained after having exposed the participants to the topic of basic logic gates with the STEAM educational approach orchestrated by the GMM, consequently, the results in Figure 8 are obtained, synthesized below:

Figure 8. Percentages of correct and incorrect answers about logic gates and their operations in the intervention group
Source: Own elaboration.

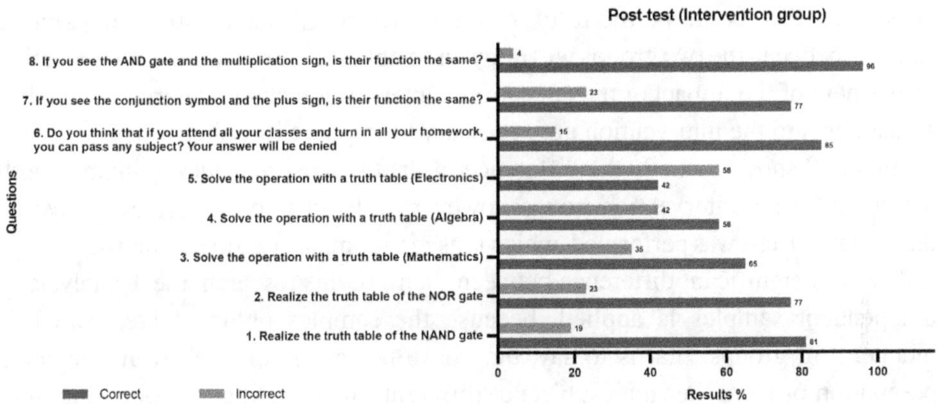

1. 81% were able to assimilate the different contexts of the NAND gate (negated AND gate), as well as its operation, while the remaining 19% had difficulties.
2. 77% were able to assimilate the different contexts of the NOR gate (negated OR gate), as well as its operation, while the remaining 23% had difficulties.
3. 65% were able to perform a complex operation ($\left(A\times\bar{B}\right)+\left(B\times\bar{A}\right)$) in the mathematical context, while 35% still had difficulties.
4. 58% were able to perform a complex operation ($\left(A\wedge\neg B\right)\vee\left(B\wedge\neg A\right)$) in the context of Boolean algebra, while the remaining 42% still had difficulties.
5. 42% managed to perform a complex operation in the context of digital electronics, while the remaining 58% still had difficulties.
6. 85% understood that the expected answer was “No”, implying that their answer was “Yes”, while the remaining 15% still had difficulties.
7. 77% were able to assimilate that the conjunction (^) is not equal to the sum, while 23% omitted the answer.
8. 96% were able to assimilate that the AND gate symbology has the same function as the multiplication operation, while the remaining 4% had difficulties.

Analysis and Synthesis of the Two Sets of Post-Test Data

Performing a post-test questionnaire analysis between a control and experimental group is crucial to evaluate the effectiveness of an educational intervention. Comparing the results between the two groups will allow us to determine whether the intervention has had a significant impact on learning and whether it has achieved the established objectives. In addition, the analysis will allow us to identify whether the results obtained are statistically significant, which will help us to determine whether the changes observed in the experimental group are really the result of the intervention or whether they could be the result of other factors. Ultimately, the comparative analysis between the two groups will allow us to obtain a more accurate and objective assessment of the impact of the educational intervention and to determine whether adjustments to the intervention are necessary to improve its effectiveness.

Figure 9 shows visually the difference of the two post-test questionnaires, both for the control and intervention groups; however, to obtain a more accurate conclusion, the student's t-test was performed, which is used to compare two means and determine if there is a significant difference between them. In this research, the T analysis of independent samples is applied, because the samples obtained are from two independent groups, that is to say, they have not been collected from the same population or from the same subject at different times, stating the following: null hypothesis (H_0) that average that has not increased, interpreting that the scores of both groups are not significant and the educational experience did not achieve the objective; alternate hypothesis (H_1) that average that has increased, indicating the

Figure 9. Visual comparison of post-test data between the control group and the intervention group
Source: Own elaboration.

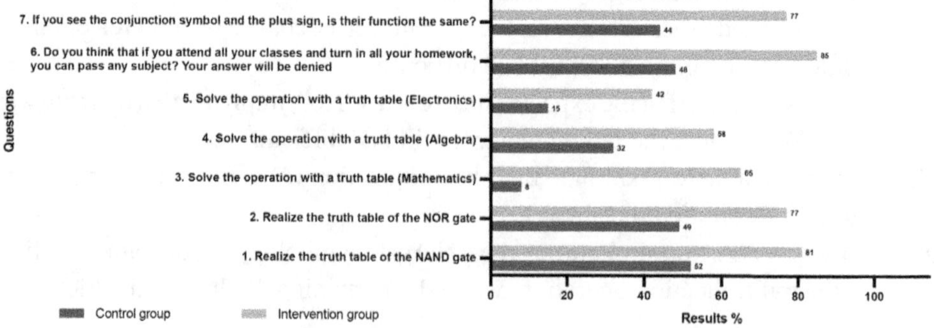

rejection of the null hypothesis and accepting the alternative hypothesis, i.e., that the difference in post-test scores was significant, achieving the objective. To perform the student's t-test, version 28 of the IBM SPSS Statistics program was used, resulting in Table 4 and Table 5.

Table 4. Group statistics

	Group	N	Mean	Std. Deviation	Std. Error Mean
Values	Control	8	.3779411765	.17545224403	.06203173576
	Intervention	8	.7259615385	.16882756671	.05968955863

Source: Own elaboration.

Table 5. Independent samples test

		Levene's Test for Equality of Variances		t-test for Equality of Means							
		F	Sig.	t	df	Significance		Mean Difference	Std. Error Difference	95% Confidence Interval of the Difference	
						One-Sided p	Two-Sided p			Lower	Upper
Values	Equal variances assumed	.104	.752	-4.043	14	<.001	.001	-.34802036200	.08608588532	-.53265622286	-.16338450114
	Equal variances not assumed			-4.043	13.979	<.001	.001	-.34802036200	.08608588532	-.53268186151	-.16335886249

Source: Own elaboration.

Student's T is negative, suggesting that there is a significant difference in the knowledge of the two groups, and in this case, the intervention group performed better than the control group. The T value is significant with a p of <0.001, indicating that the probability that this difference is due to chance is very low. The 95% confidence interval ($\alpha = 0.05$) indicates that the true value of difference between the groups is -0.532656 (lower) and -0.163385 (upper), suggesting that the intervention group had a significant increase in knowledge compared to the control group, thus accepting the alternative hypothesis.

CONCLUSION

During the research we found that the GMM effectively and efficiently orchestrates educational activities with three or more STEAM disciplines using the three stages that comprise it. The HTL, although it may cause some confusion because these are linear and the one presented has multiple paths due to the context of each discipline, it should not be forgotten that there are Real Learning Trajectory, which are those that are created from the learning channel, prior knowledge, preferences and decisions of students, which can be seen as the actual path that the student travels, what is intended with the HTL presented, is to offer more than one path to achieve the educational objective.

On the other hand, it is possible to integrate both hardware and software to favor educational activities of STEAM nature, it is emphasized that the role of technology in education is only supportive and not the main reason for its use. Furthermore, that this type of technological resources allows to scale and deepen the knowledge as much as required to obtain the expected objectives.

Instructional design is also a very important resource to identify and apply knowledge in the different areas of STEAM, as well as the control of time, materials, exercises, and evaluations. Therefore, the GMM does not intend that the teacher changes his or her way of teaching, only that he or she has the appropriate tools to guide students in this type of activities that require the presentation of a topic in different disciplines, not to mention that it allows effective and efficient guidance on the use of technology.

Future Research

It is important that future research contemplate the use of statistics to check if there has been a significant change between the same group (pre and post-test) and different groups (post-test), applying it to different educational levels, this will strengthen the proposed instructional designs and in the case of the GMM, will allow better organizing the information for classroom instruction, hoping to improve the results. Likewise, it is recommended that researchers guide teachers in the preparation of STEAM activities, since teachers are the first line in education.

REFERENCES

Alba, C. (2019). Diseño Universal para el Aprendizaje: un modelo teórico-práctico para una educación inclusiva de calidad. *Participación educativa*, *6*(9), 55-66.

Areljung, S., & Günther-Hanssen, A. (2022). STEAM education: An opportunity to transcend gender and disciplinary norms in early childhood? *Contemporary Issues in Early Childhood*, *23*(4), 500–503. doi:10.1177/14639491211051434

Arufe, V., Sanmiguel-Rodríguez, A., Ramos, O. & Navarro-Patón, R. (2022). Can Gamification Influence the Academic Performance of Students? *Sustainability, 14*(9), 5115. doi:10.3390/su14095115

Avalos, M. & Castellanos, D. (2022). La disrupción de la educación presencial a la modalidad virtual, una propuesta de diseño instruccional-tecnopedagógico. *Revista Hamut́ay*, *9* (2), 27-36. doi:10.21503/hamu.v9i2.2429

Avdeenko, A., Sabirova, F., & Konyushenko, S. (2021). STEM Education: Key Challenges in the Common Practice. [IJWLTT]. *International Journal of Web-Based Learning and Teaching Technologies*, *16*(6), 1–11. doi:10.4018/IJWLTT.293281

Catlin, D., Kandlhofer, M., & Holmquist, S. (2018). EduRobot Taxonomy: A Provisional Schema for Classifying Educational Robots. In *Robotics in Education 2018*. http://rie2018.info/

Chang, C., Debra, C., & Chang, W. (2019). Research on Immersion for Learning Using Virtual Reality, Augmented Reality and Mixed Reality. *Enfance*, *3*(3), 413–426. doi:10.3917/enf2.193.0413

Chung, C. C., Huang, S. L., Cheng, Y. M., & Lou, S.-J. (2022). Using an iSTEAM project-based learning model for technology senior high school students: Design, development, and evaluation. *International Journal of Technology and Design Education*, *32*(2), 905–941. doi:10.100710798-020-09643-5

Clark, D., Sengupta, P., Brady, C., Martinez-Garza, M. M., & Killingsworth, S. S. (2015). Disciplinary integration of digital games for science learning. *International Journal of STEM Education*, *2*(1), 2. doi:10.118640594-014-0014-4

Conde, M., Rodríguez-Sedano, F., Fernández-Llamas, C., Gonçalves, J., Lima, J., & García-Peñalvo, F. (2021). Fostering STEAM through challenge-based learning, robotics, and physical devices: A systematic mapping literature review. *Computer Applications in Engineering Education*, *29*(1), 46–65. doi:10.1002/cae.22354

Flores-Nicolás, M., & Reyes, M. (2022). The Graded Multidisciplinary Model: Fostering instructional design for activity development in STEM/STEAM education. In *2022 IEEE Mexican International Conference on Computer Science (ENC)* (pp. 1-7). IEEE. 10.1109/ENC56672.2022.9882917

Greca, I., Ortiz-Revilla, J. & Arriassecq, I. (2021). Diseño y evaluación de una secuencia de enseñanza-aprendizaje STEAM para Educación Primaria. *Revista Eureka sobre Enseñanza y Divulgación de las Ciencias*, *18*(1), 1-19. doi:10.25267/Rev_Eureka_ensen_divulg_cienc.2021.v18.i1.1802

Huffman, D., Thomas, K., & Basham, J. (2020). The Transdisciplinary Nature of STEAM Education: Integrating STEAM in Pre-Service Teacher Education. In K. Thomas & D. Huffman (Eds.), *Challenges and Opportunities for Transforming From STEM to STEAM Education* (pp. 221–237). IGI Global. doi:10.4018/978-1-7998-2517-3.ch009

Ivars, P., Fernández, C., & Llinares, S. (2020). Uso de una trayectoria hipotética de aprendizaje para proponer actividades de instrucción. *Enseñanza de las Ciencias, 38*(3), 105–124. doi:10.5565/rev/ensciencias.2947

Jackson, H., Basham, J., Thomas, K., & Hunt, C. (2020). Designing STEAM Learning Environments. In K. Thomas & D. Huffman (Eds.), *Challenges and Opportunities for Transforming From STEM to STEAM Education* (pp. 1–23). IGI Global. doi:10.4018/978-1-7998-2517-3.ch001

Jensen, E., & Hanghøj, T. (2020). What's the math in Minecraft? A Design-Based Study of Students' Perspectives and Mathematical Experiences Across game and School Domains. *Electronic Journal of e-Learning, 18*(3), 261–274. doi:10.34190/EJEL.20.18.3.005

Juškevičienė, A., Dagienė, V., & Dolgopolovas, V. (2021). Integrated activities in STEM environment: Methodology and implementation practice. *Computer Applications in Engineering Education, 29*(1), 209–228. doi:10.1002/cae.22324

Krüger, M., & Chiappe, A. (2021). 21st-century skills and their relationshio to STEAM learning environments: A review. [RED]. *Revista de Educación a Distancia, 21*(68). doi:10.6018/red.470461

Li, Y., Wang, K., Xiao, Y., & Froyd, J. E. (2020). Research and trends in STEM education: A systematic review of journal publications. *International Journal of STEM Education, 7*(1), 11. doi:10.118640594-020-00207-6

Li, Y., Xiao, Y., Wang, K., Zhang, N., Pang, Y., Wang, R., Qi, C., Yuan, Z., Xu, J., Nite, S. B., & Star, J. R. (2022). A systematic review of high impact empirical studies in STEM education. *International Journal of STEM Education, 9*(1), 72. doi:10.118640594-022-00389-1

Liston, M., Morrin, A. M., Furlong, T., & Griffin, L. (2022). Integrating Data Science and the Internet of Things Into Science, Technology, Engineering, Arts, and Mathematics Education Through the Use of New and Emerging Technologies. In *Frontiers in Education* (p. 348). Frontiers. doi:10.3389/feduc.2022.757866

Lorenzo-Lledó, A. (2023). El modelo educativo inclusivo desde el Diseño Universal para el Aprendizaje (DUA). *Atención a la Diversidad en los Centros Educativos.* [Online]. Disponible en: https://rua.ua.es/dspace/bitstream/10045/131375/1/El-modelo-educativo-inclusivo-desde-el-Diseno-Universal-para-el-Aprendizaje-DUA.pdf

MacDonald, A., Wise, K., Tregloan, K., Fountain, W., Wallis, L., & Holmstrom, N. (2020). Designing STEAM education: Fostering relationality through design-led disruption. *International Journal of Art & Design Education, 39*(1), 227–241. doi:10.1111/jade.12258

Margot, K., & Kettler, T. (2019). Teachers' perception of STEM integration and education: A systematic literature review. *International Journal of STEM Education, 6*(1), 2. doi:10.118640594-018-0151-2

Martínez, P. (2006). El método de estudio de caso: estrategia metodológica de la investigación científica. *Pensamiento & gestión*, (20), 165-193. https://www.redalyc.org/pdf/646/64602005.pdf

McLure, F., Tang, K., & Williams, P. (2022). What do integrated STEM projects look like in middle school and high school classrooms? A systematic literature review of empirical studies of iSTEM projects. *International Journal of STEM Education, 9*(1), 73. doi:10.118640594-022-00390-8

Miller, T. (2018). Developing numeracy skills using interactive technology in a play-based learning environment. *International Journal of STEM Education, 5*(1), 39. doi:10.118640594-018-0135-2 PMID:30631729

Nasri, N., Rahimi, N., Nasri, N. & Talib, M. (2021). A Comparison Study between Universal Design for Learning-Multiple Intelligence (UDL-MI) Oriented STEM Program and Traditional STEM Program for Inclusive Education. *Sustainability, 13*(2), 554. MDPI AG. doi:10.3390/su13020554

Ohnishi, Y., Honda, K., Nishioka, R., Mori, S. & Kawada, K. (2017). Robotics programming learning for elementary and junior high school students. *Journal of robotics and mechatronics, 29*(6), 992-998. doi:10.20965/jrm.2017.p0992

Peterson, B., & Hipple, B. (2020). Formative Assessment in Hands-On STEM Education. In C. Martin, D. Polly, & R. Lambert (Eds.), *Handbook of Research on Formative Assessment in Pre-K Through Elementary Classrooms* (pp. 1–29). IGI Global. doi:10.4018/978-1-7998-0323-2.ch009

Quigley, C., Herro, D., & Jamil, F. (2017). Developing a conceptual model of STEAM teaching practices. *School Science and Mathematics, 117*(1-2), 1–12. doi:10.1111sm.12201

Schuelka, M., & Thyrring, T. (2022). On the question of educational purpose: Complex educational systems analysis for inclusion. *International Journal of Inclusive Education, 26*(5), 448–465. doi:10.1080/13603116.2019.1698062

Tekerek, M., & Tekerek, B. (2018). Integrated instructional material and development processes. *Turkish Journal of Education, 7*(3), 156–168. doi:10.19128/turje.362491

Urgiles, B., Tixi, K., Allauca, M., & Fonseca, G. (2021). Systematic review of the use of STEAM in higher education. *MktDESCUBRE, 1*(17), 85–94. doi:10.36779/mktdescubre.v1i17.580

Viskupic, K., Earl, B., & Shadle, S. (2022). Adapting the CACAO model to support higher education STEM teaching reform. *International Journal of STEM Education, 9*(1), 6. doi:10.118640594-021-00325-9

Wang, L. H., Chen, B., Hwang, G. J., Guan, J.-Q., & Wang, Y.-Q. (2022). Effects of digital game-based STEM education on students' learning achievement: A meta-analysis. *International Journal of STEM Education, 9*(1), 26. doi:10.118640594-022-00344-0

Yin, B., & Yuan, C. H. (2021). Precision teaching and learning performance in a blended learning environment. *Frontiers in Psychology, 141*(12), 631125. doi:10.3389/fpsyg.2021.631125 PMID:33613404

ADDITIONAL READING

ISTE. (2023). *International Society for Technology in Education*. ISTE. https://www.iste.org/iste-standards

Justin. (2023). *Educommunity.minecraft.net*. EduCommunity. https://educommunity.minecraft.net/hc/en-us/articles/360047556591-System-Requirements

Microsoft Learn. (2023). *Minecraft learning paths (Training)*. Microsoft. https://learn.microsoft.com/es-es/training/browse/?terms=minecraft

Mojan (2023). *Minecraft.net*. EduCommity. https://education.minecraft.net/es-es/discover/what-is-minecraft

Compilation of References

Aazam, M. (2018). Fog computing for healthcare: A review of current trends and future directions. *Journal of Medical Systems*, *42*(7), 1–13.

Aazam, S. (2018). Fog computing: Issues and challenges in security and privacy. *Journal of Network and Computer Applications*, *116*, 1–15.

Aboubakar, M., Kellil, M., & Roux, P. (2022). A review of IoT network management: Current status and perspectives. *Journal of King Saud University - Computer and Information Sciences*, *34*(7), 4163–4176. doi:10.1016/j.jksuci.2021.03.006

Afonasova, M. A., Panfilova, E. E., Galichkina, M. A., & Slusarczyk, B. (2019). Digitalization in economy and innovation: The effect on social and economic processes. *Polish Journal of Management Studies*, *19*(2), 23–32. doi:10.17512/pjms.2019.19.2.02

Agarwall, H., Das, C. P., & Swain, R. K. (2022). *Does artificial intelligence influence the operational performance of companies? A study*. Atlantis Highlights in Social Sciences, Education and Humanities., doi:10.2991/ahsseh.k.220105.008

Aggarwal, D., & Maurer, U. (2016). Breaking RSA Generically Is Equivalent to Factoring. IEEE Transactions on Information Theory, 62(11), 6251–6259. doi:10.1109/TIT.2016.2594197

Aggarwal, C., & Subbian, K. (2014). Evolutionary network analysis. *ACM Computing Surveys*, *47*(1), 1–36. doi:10.1145/2601412

Ahmed, W. A. H., MacCarthy, B. L., & Treiblmaier, H. (2022). Why, where and how are organizations using blockchain in their supply chains? Motivations, application areas and contingency factors. *International Journal of Operations & Production Management*, *42*(12), 1995–2028. doi:10.1108/IJOPM-12-2021-0805

Ahmetoglu, S., Che Cob, Z., & Ali, N. A. (2022). A systematic review of Internet of Things adoption in organizations: Taxonomy, benefits, challenges and critical factors. *Applied Sciences (Basel, Switzerland)*, *12*(9), 4117. doi:10.3390/app12094117

Akhtar, P., Khan, Z., Tarba, S., & Jayawickrama, U. (2017). The Internet of Things, dynamic data and information processing capabilities, and operational agility. *Technological Forecasting and Social Change*, *136*, 307–316. doi:10.1016/j.techfore.2017.04.023

Alabadi, M., & Çelik, Y. (2020). Anomaly detection for cyber-security based on convolutional neural network: A survey. *2020 International Congress on Human-Computer Interaction, Optimization and Robotic Applications (HORA)*. 10.1109/HORA49412.2020.9152899

Alabadi, M., Habbal, A., & Wei, X. (2022). Industrial internet of things: Requirements, architecture, challenges, and future research directions. *IEEE Access : Practical Innovations, Open Solutions*, *10*, 66374–66400. doi:10.1109/ACCESS.2022.3185049

Alba, C. (2019). Diseño Universal para el Aprendizaje: un modelo teórico-práctico para una educación inclusiva de calidad. *Participación educativa*, *6*(9), 55-66.

Al-Qaseemi, S. A., Almulhim, H. A., Almulhim, M. F., & Chaudhry, S. R. (2016, December). IoT architecture challenges and issues: Lack of standardization. In *2016 Future technologies conference (FTC)* (pp. 731–738). IEEE. doi:10.1109/FTC.2016.7821686

Alvee, S. R. B., Ahn, B., Kim, T., Su, Y., Youn, Y.-W., & Ryu, M.-H. (2021). Ransomware Attack Modeling and Artificial Intelligence-Based Ransomware Detection for Digital Substations. In *2021 6th IEEE Workshop on the Electronic Grid (eGRID). 2021 6th IEEE Workshop on the Electronic Grid (eGRID)*. IEEE. 10.1109/eGRID52793.2021.9662158

Ane, T., Billah, M., & Nepa, T. (2020). Performance of Internet of things (IoT) potential applications in education. *Bangladesh Journal of Multidisciplinary Scientific Research*, *2*(2), 10–16. doi:10.46281/bjmsr.v2i2.653

Anshari, M., Syafrudin, M., & Fitriyani, N. L. (2022). Fourth Industrial Revolution between Knowledge Management and Digital Humanities. *Information (Basel)*, *13*(6), 292. doi:10.3390/info13060292

Anute, N., Paliwal, M., Patel, M., & Kandale, N. (2021). Impact of artificial intelligence and machine learning on business operations. *Journal of Management Research and Analysis*, *8*(2), 69–74. doi:10.18231/j.jmra.2021.015

Araya, A. A. (1995, February). Questioning ubiquitous computing. In *Proceedings of the 1995 ACM 23rd annual conference on Computer science* (pp. 230–237). 10.1145/259526.259560

Areljung, S., & Günther-Hanssen, A. (2022). STEAM education: An opportunity to transcend gender and disciplinary norms in early childhood? *Contemporary Issues in Early Childhood*, *23*(4), 500–503. doi:10.1177/14639491211051434

Arufe, V., Sanmiguel-Rodríguez, A., Ramos, O. & Navarro-Patón, R. (2022). Can Gamification Influence the Academic Performance of Students? *Sustainability*, *14*(9), 5115. doi:10.3390/su14095115

Asbaş, C., & Erdem Tuzlukaya, Ş. (2021). Dijital çağda sosyal ağlar ve sosyal girişimcilik. In D. Hıdıroğlu (Ed.), *Dijital Çağda Kurumsal Yönetişim, Sosyal Sorumluluk ve Sosyal Girişimcilik* (pp. 395–418). Artikel Akademi.

Asbaş, C., & Erdem Tuzlukaya, Ş. (2022). Cyberattack and cyberwarfare strategies for businesses. In F. Özsungur (Ed.), *Conflict Management in Digital Business: New Strategy and Approach* (pp. 303–329). Emerald. doi:10.1108/978-1-80262-773-220221027

Asbaş, C., & Erdem Tuzlukaya, Ş. (2023). Cyberwarfare: War activities in cyberspace. In F. Özsungur (Ed.), *Handbook of Research on War Policies, Strategies, and Cyber Wars* (pp. 128–145). IGI Global. doi:10.4018/978-1-6684-6741-1.ch007

ASD. (2023). *Autism Spectrum Disorder, Signs and Symptoms of Autism Spectrum Disorder*. CDC. https://www.cdc.gov/ncbddd/autism/signs.html

Ashton, K. (2009). That 'Internet of things' thing. *RFID Journal*, *22*(7), 97–114.

Asjad, M., Alam, A., & Hasan, F. (2018). A comparative study of classifier techniques for lift index data analysis. *Benchmarking*, *25*(2), 632–641. doi:10.1108/BIJ-09-2016-0137

Avalos, M. & Castellanos, D. (2022). La disrupción de la educación presencial a la modalidad virtual, una propuesta de diseño instruccional-tecnopedagógico. *Revista Hamut'ay*, *9* (2), 27-36. doi:10.21503/hamu.v9i2.2429

Avdeenko, A., Sabirova, F., & Konyushenko, S. (2021). STEM Education: Key Challenges in the Common Practice. [IJWLTT]. *International Journal of Web-Based Learning and Teaching Technologies*, *16*(6), 1–11. doi:10.4018/IJWLTT.293281

Azimi, R., & Alcaraz, C. (2020). A blockchain-based smart contract for secure data sharing in healthcare. *Journal of Medical Systems*, *44*(1), 14. doi:10.100710916-019-1521-9

Bag, S., Rahman, M.S., Gupta, S., & Wood, L.C. (2022). Understanding and predicting the determinants of blockchain technology adoption and SMEs' performance. *The International Journal of Logistics Management*. doi:10.1108/IJLM-01-2022-0017

Barreno A. A. & Monar N. M. (2023). *Parametric design and production in Manufacturing Laboratory (Fablab), as an alternative to reduce the emission of carbon dioxide*. Case study. Technical University of Ambato. Center for Studies in Design and Communication.

Barrientos-Avendaño, E., Areniz-Arevalo, Y., Coronel-Rojas, L. A., Cuesta-Quintero, F., & Rico-Bautista, D. (2020). Modelo de incursión en la industria 4.0 aplicado a la compañía alimenticia tu pan gourmet SAS: Estrategia para el renacer en la pandemia ocasionada por COVID-19 (SARS-CoV-2). *Revista Ibérica de Sistemas e Tecnologias de Informação*, (E34), 436–449.

Becerra, W. J. A., Luna, A. R. L., & Aponte, G. C. V. (2021). Efectos organizacionales durante la pandemia covid-19: El caso de conamovidi y los comedores autogestionarios. *Revista de Sociologia (Lima)*, (33), 61–76. doi:10.15381/rsoc.n33.21795

Bekara, C. (2014). Security issues and challenges for the IoT-based smart grid. *Procedia Computer Science*, *34*, 532–537. doi:10.1016/j.procs.2014.07.064

Bellis, M. (2023). *Biography of James Watt, Inventor of the Modern Steam Engine.* ThoughtCo. https://www.thoughtco.com/james-watt-inventor-of-the-modern-steam-engine-1992685

Bernstein, D. J., & Lange, T. (2016). *SafeCurves: Choosing safe curves for elliptic-curve cryptography.* https://safecurves.cr.yp.to/

Bhubalan, K., Tamothran, A. M., Kee, S. H., Foong, S. Y., Lam, S. S., Ganeson, K., Vigneswari, S., Amirul, A.-A., & Ramakrishna, S. (2022). Leveraging blockchain concepts as watermarkers of plastics for sustainable waste management in progressing circular economy. *Environmental Research, 213*, 113631. doi:10.1016/j.envres.2022.113631 PMID:35714685

Blanc, R., Ratto, D., Cettour, W., & Lepratte, L. (2021). *Modelos de madurez e implementación en industria 4.0 análisis de alternativas y nivel de implementación.* Academic Press.

Blanchet, M. J. O. T. (2016). Industrie 4.0: Nouvelle donne industrielle, nouveau modèle économique. *Outre-Terre*, *46*(1), 62–85. doi:10.3917/oute1.046.0062

Blos, M. F., da Silva, R. M., & Wee, H. M. (2018). A framework for designing supply chain disruptions management considering productive systems and carrier viewpoints. *International Journal of Production Research*, *56*(15), 5045–5061. doi:10.1080/00207543.2018.1442943

Bondy, M., Roth, S., & Sager, L. (2020). Crime Is in the Air: The Contemporaneous Relationship between Air Pollution and Crime. *Journal of the Association of Environmental and Resource Economists*, *7*(3), 555–585. doi:10.1086/707127

Bonneau, J., Clark, J., & Goldfeder, S. (2015). On bitcoin as a public randomness source. IACR Cryptol., 1015.

Bourdon, K. H., Boyd, J. H., Rae, D. S., Burns, B. J., Thompson, J. W., & Locke, B. Z. (1988). Gender differences in phobias: Results of a ECA community survey. *Journal of Anxiety Disorders*, *2*(3), 227–241. doi:10.1016/0887-6185(88)90004-7

Britannica. (2023). *Industrial Revolution.* Britannica. https://www.britannica.com/event/Industrial-Revolution.

Brown. (2016). *SECG SEC 1: Elliptic Curve Cryptography (Version 2.0).* https://www.secg.org/sec1-v2.pdf

Brynjolfsson, E., Rock, D., & Syverson, C. (2017). *Artificial intelligence and the modern productivity paradox: A clash of expectations and statistics* (Vol. 24001). NBER Working Paper. doi:10.3386/w24001

Buckingham-Hatfield, S., & Evans, B. (1996). Achieving sustainability through environmental planning. In S. Buckingham-Hatfield & B. Evans (Eds.), Environmental Planning and Sustainability. John Wiley & Sons.

Bughin, J., Hazan, E., Lund, S., Dahlström, P., Wiesinger, A., & Subramaniam, A. (2018). *Skill Shift: Automation and The Future of the Workforce*. McKinsey Global Institute.

Burgess, M., Adar, E., & Cafarella, M. (2016). Link-prediction enhanced consensus clustering for complex networks. *PLoS One*, *11*(5), e0153384. Advance online publication. doi:10.1371/journal.pone.0153384 PMID:27203750

Butt, J. (2020). Exploring the interrelationship between additive manufacturing and Industry 4.0. *Designs*, *4*(2), 13. doi:10.3390/designs4020013

Buyya, R. (2018). Fog computing: A survey of enabling technologies, applications, and issues. *IEEE Internet of Things Journal*, *5*(2), 524–545.

Caballero, H. A. (2013). *Recategorization and enhancement of the Carpentry Workshop of the Puerto Belgrano Naval Arsenal. University of the Fraternity of Groups St. Thomas Aquinas.* Faculty of Engineering.

Cabrera-Ramírez, S., & Cepeda-Retana, J. (2022). La epistemología, guía para el conocimiento científico. *Portal de la Ciencia*, *3*(2), 123–133. doi:10.51247/pdlc.v3i2.317

Can Saglam, Y., Sezen, B., & Çankaya, S. Y. (2020). The inhibitors of risk information sharing in the supply chain: A multiple case study in Turkey. *Journal of Contingencies and Crisis Management*, *28*(1), 19–29. doi:10.1111/1468-5973.12285

Çankaya, S. Y., & Sezen, B. (2020). Industry 4.0 and Sustainability. In Handbook of Research on Creating Sustainable Value in the Global Economy (pp. 67-84). IGI Global.

Cano-Orón, L., & Llorca, G. (2018). La configuración invisible del espacio digital: el caso de Facebook. *Observatorio (OBS*), 12*(3), 106-121.

Cao, M. (2019). Fog computing for smart grid: A survey. *IEEE Transactions on Industrial Informatics*, *15*(1), 104–118.

Cardona, C., Restrepo, D., & Ovalle, D. (2007). Sistema Experto Difuso para Determinar Perfiles Criminológicos basado en el Test de Lüscher y Variables Socio-Criminológicas. *Revista Avances en Sistemas e Informática, 4*(2), 97-105. Recuperado de: https://www.redalyc.org/articulo.oa?id=133116858012

Carvalho, de B. L., Bassanezi, R. C., & Lodwick, W. A. (2017). *A first course in fuzzy logic, fuzzy dynamical systems, and Biomathematics Theory and Applications*. Springer Berlin Heidelberg.

Catlin, D., Kandlhofer, M., & Holmquist, S. (2018). EduRobot Taxonomy: A Provisional Schema for Classifying Educational Robots. In *Robotics in Education 2018.* http://rie2018.info/

Chang, C., Debra, C., & Chang, W. (2019). Research on Immersion for Learning Using Virtual Reality, Augmented Reality and Mixed Reality. *Enfance*, *3*(3), 413–426. doi:10.3917/enf2.193.0413

Chen, T., Li, X., Luo, X., & Zhang, X. (2017). Under-optimized smart contracts devour your money. *Proceedings of 24th International Conference on Software Analysis, Evolution and Reengineering*, 442–446. 10.1109/SANER.2017.7884650

Chen, D., Esperança, J. P., & Wang, S. (2022). The impact of artificial intelligence on firm performance: An application of the resource-based view to e-commerce firms. *Frontiers in Psychology, 13*, 1–14. doi:10.3389/fpsyg.2022.884830 PMID:35465474

Chen, S. (2020). Energy-efficient fog computing: A review of research challenges and future directions. *IEEE Network, 34*(2), 22–28.

Chen, S., Yang, H., Gao, X., Lin, W., & Zhang, H. (2020). Energy-efficient fog computing: A review of research challenges and future directions. *Journal of Parallel and Distributed Computing, 140*, 99–112.

Choi, T. M., Kumar, S., Yue, X., & Chan, H. L. (2022). Disruptive technologies and operations management in the Industry 4.0 era and beyond. *Production and Operations Management, 31*(1), 9–31. doi:10.1111/poms.13622

Chung, C. C., Huang, S. L., Cheng, Y. M., & Lou, S.-J. (2022). Using an iSTEAM project-based learning model for technology senior high school students: Design, development, and evaluation. *International Journal of Technology and Design Education, 32*(2), 905–941. doi:10.100710798-020-09643-5

Cisneros, J. L. (2020). Violencia y criminalidad: El caso del Municipio de Chalco, Estado de México. *SAPIENTIAE: Revista de Ciencias Sociais, Humanas e Engenharias, 5*(2), 272-285. Recuperado de: https://www.redalyc.org/articulo.oa?id=572762223002

Clark, D., Sengupta, P., Brady, C., Martinez-Garza, M. M., & Killingsworth, S. S. (2015). Disciplinary integration of digital games for science learning. *International Journal of STEM Education, 2*(1), 2. doi:10.118640594-014-0014-4

Clauset, A., Moore, C., & Newman, M. E. (2008). Hierarchical structure and the prediction of missing links in networks. *Nature, 453*(7191), 98–101. doi:10.1038/nature06830 PMID:18451861

Clavijo-Quintero, D. P. (2022). *Análisis de la Industria 4.0 En América Latina: Caso México y Colombia*. Academic Press.

Conde, M., Rodríguez-Sedano, F., Fernández-Llamas, C., Gonçalves, J., Lima, J., & García-Peñalvo, F. (2021). Fostering STEAM through challenge-based learning, robotics, and physical devices: A systematic mapping literature review. *Computer Applications in Engineering Education, 29*(1), 46–65. doi:10.1002/cae.22354

Condori, V. C. (2018). Reconditioning of a two-stage industrial sander for the company "MAEXA S.R.L." Universidad Mayor de San Andrés. Faculty of Technology. Electromechanical Career.

Constantinides, P., Henfridsson, O., & Parker, G. G. (2018). Introduction – Platforms and infrastructures in the digital age. *Information Systems Research, 29*(2), 381–400. doi:10.1287/isre.2018.0794

Contreras, A. J. A. (2021). *Design and manufacture of a CNC milling machine. University of Jaén. Higher Polytechnic School of Linares*. Department of Mechanical Engineering.

Coyoli-Pereyra, D. M. (2022). *La calidad de vida a través de los sentimientos en las revoluciones industriales*.

Cremer, F., Sheehan, B., Fortmann, M., Kia, A. N., Mullins, M., Murphy, F., & Materne, S. (2022). Cyber risk and cybersecurity: a systematic review of data availability. In The Geneva Papers on Risk and Insurance - Issues and Practice (Vol. 47, Issue 3, pp. 698–736). Springer Science and Business Media LLC. doi:10.105741288-022-00266-6

Cunha, P. R. D., Soja, P., & Themistocleous, M. (2021). Blockchain for development: A guiding framework. *Information Technology for Development*, *27*(3), 417–438. doi:10.1080/02681102.2021.1935453

da Silva, C. F., & Moro, S. (2021). Blockchain technology as an enabler of consumer trust: A text mining literature analysis. *Telematics and Informatics*, *60*, 101593. doi:10.1016/j.tele.2021.101593

Da Xu, L., He, W., & Li, S. (2014). Internet of things in industries: A survey. *IEEE Transactions on Industrial Informatics*, *10*(4), 2233–2243. doi:10.1109/TII.2014.2300753

Damanpour, F., & Aravind, D. (2012). Managerial innovation: Conceptions, processes, and antecedents. *Management and Organization Review*, *8*(2), 423–454. doi:10.1111/j.1740-8784.2011.00233.x

Daza, M. U. R. & Luengas, Z. S. (2016). *CNC Router Applied to the Development of PCB's*. National Polytechnic Institute. School of Mechanical and Electrical Engineering Professional Unit Adolfo López Mateos.

Delic, M., Eyers, D. R., & Mikulic, J. (2019). Additive manufacturing: Empirical evidence for supply chain integration and performance from the automotive industry. *Supply Chain Management*, *24*(5), 604–621. doi:10.1108/SCM-12-2017-0406

Demir, C., & Ozcan, S. E. (2022). Do air quality and green space reduce propensity to crime?: an empirical investigation of OECD countries. Management of Environmental Quality. doi:10.1108/MEQ-08-2022-0235

Devang, V., Chintan, S., Gunjan, T., & Krupa, R. (2019). Applications of artificial intelligence in marketing. *Annals of Dunarea de Jos University of Galati Fascicle I. Economics and Applied Informatics*, *25*(1), 28–36. doi:10.35219/eai158404094

Dunlavy, D. M., Kolda, T. G., & Acar, E. (2011). Temporal link prediction using matrix and tensor factorizations. *ACM Transactions on Knowledge Discovery from Data*, *5*(2), 1–27. doi:10.1145/1921632.1921636

Dursun, E., Ulker, Y., & Gunalay, Y. (2022). Blockchain's potential for waste management in textile industry. Management of Environmental Quality. doi:10.1108/MEQ-03-2022-0085

Elhusseiny, H. M., & Crispim, J. (2022). SMEs, Barriers and Opportunities on adopting Industry 4.0: A Review. *Procedia Computer Science*, *196*, 864–871. doi:10.1016/j.procs.2021.12.086

Farahani, B., Firouzi, F., Chang, V., Badaroglu, M., & Constant, N. (n.d.). *Towards fog-driven IoT eHealth: Promises and challenges of IoT in medicine and healthcare.* Future20. doi:10.1016/j.future.2017.04.036

Faurot, K., Dellinger, D., Eatough, A., & Parkhurst, S. (1992). *The identity of Mexican sign as a language.* SIL International.

Ferrer, A. (2022). *Historia de la globalización II: La Revolución Industrial y el segundo orden mundial.* Fondo de Cultura Económica Argentina.

Flores Payán, L., & Refugio Vallejo, J. (2015). Evaluación de políticas y programas sociales mediante lógica difusa. *Nóesis. Revista de Ciencias Sociales y Humanidades, 24*(1), 83–114. doi:10.20983/noesis.2015.1.3

Flores-Nicolás, M., & Reyes, M. (2022). The Graded Multidisciplinary Model: Fostering instructional design for activity development in STEM/STEAM education. In *2022 IEEE Mexican International Conference on Computer Science (ENC)* (pp. 1-7). IEEE. 10.1109/ENC56672.2022.9882917

Floridi, L., Cowls, J., Beltrametti, M., Chatila, R., Chazerand, P., Dignum, V., Luetge, C., Madelin, R., Pagallo, U., Rossi, F., Schafer, B., Valcke, P., & Vayena, E. (2018). AI4People–An ethical framework for a good AI society: Opportunities, risks, principles, and recommendations. *Minds and Machines, 28*(4), 689–707. doi:10.100711023-018-9482-5 PMID:30930541

Fosso Wamba, S., & Guthrie, C. (2020). The impact of blockchain adoption on competitive performance: The mediating role of process and relational innovation. *Logistique & Management, 28*(1), 88–96. doi:10.1080/12507970.2019.1679046

Franco, M. S., Valdez P. M. (2022). *Principios básicos de cartografía y cartografía automatizada.* UNAM.

Freilich, J. D., & Newman, G. R. (2017). Situational crime prevention. Oxford research encyclopedia of criminology and criminal justice. doi:10.1093/acrefore/9780190264079.013.3

Front labeling of food and beverages. (2021, October 5). gob.mx. https://www.gob.mx/promosalud/acciones-y-programas/etiquetado-de-alimentos

Galiautdinov, R., & Mkrttchian, V. (2019b). Brain machine interface – for Avatar Control & Estimation in Educational purposes Based on Neural AI plugs: Theoretical and Methodological Aspects. In Avatar-Based Control, Estimation, Communications, and Development of Neuron Multi-Functional Technology Platforms (pp. 320–344). IGI Global. doi:10.1016/j.comnet.2021.108011

Galiautdinov, R. (2020). *Brain machine interface: the accurate interpretation of neurotransmitters' signals targeting the muscles. International Journal of Applied Research in Bioinformatics.* doi:10.4018/IJARB.2020010102

Galiautdinov, R., & Mkrttchian, V. (2019a). Math model of neuron and nervous system research, based on AI constructor creating virtual neural circuits: Theoretical and Methodological Aspects. In V. Mkrttchian, E. Aleshina, & L. Gamidullaeva (Eds.), *Avatar-Based Control, Estimation, Communications, and Development of Neuron Multi-Functional Technology Platforms* (pp. 320–344). IGI Global. doi:10.4018/978-1-7998-1581-5.ch015

García, R. C., & Oliva, M. A. (2021). La transformación digital de los medios de pago: Impactos de la covid-19. In Managing the future: Challenges and Proposals for Post-pandemic Society (pp. 243-253). Universitat Rovira i Virgili.

Garcia, M. I. C., Lagos, L. J. G., Urrego, P. L. F., & Delgado, P. P. Y. (2009). Design and implementation of a CNC control to create models and sculptures in third dimension from a CAD design. University of San Buenaventura. Faculty of Engineering. Mechatronics Program. *Bogotá D., C*, 190.

Gaur, N. (2020). Blockchain challenges in adoption. *Managerial Finance*, *46*(6), 849–858. doi:10.1108/MF-07-2019-0328

Geografía (INEGI). (2022, January 1). *Incidencia delictiva.* Encuestas En Establecimientos. Especiales. Encuesta Nacional de Victimización de Empresas. ENVE, Encuestas En Hogares. Encuesta Nacional de Victimización Y Percepción Sobre Seguridad Pública. ENVIPE. https://www.inegi.org.mx/temas/incidencia/

George, R. P., Peterson, B. L., Yaros, O., Beam, D. L., Dibbell, J. M., & Moore, R. C. (2019). Blockchain for business. *Journal of Investment Compliance*, *20*(1), 17–21. doi:10.1108/JOIC-01-2019-0001

Ghimire, T., Joshi, A., Sen, S., Kapruan, C., Chadha, U., & Selvaraj, S. K. (2022). Blockchain in additive manufacturing processes: Recent trends & its future possibilities. *Materials Today: Proceedings*, *50*, 2170–2180. doi:10.1016/j.matpr.2021.09.444

Gokceoglu, C., & Sezer, E. (2012). Soft computing modeling in landslide susceptibility assessment. In *Terrigenous mass movements* (pp. 51–90). Springer. doi:10.1007/978-3-642-25495-6_3

Gómez-Chabla, R., Real-Avilés, K., Morán, C., Grijalva, P., & Recalde, T. (2018, December). IoT applications in agriculture: A systematic literature review. In *ICT for Agriculture and Environment: Second International Conference, CITAMA 2019, Guayaquil, Ecuador, January 22-25, 2019, Proceedings* (pp. 68-76). Cham: Springer International Publishing. 10.1007/978-3-030-10728-4_8

Gómez, R. R. (2022). Internet de las cosas, salud pública y pandemias. *Salud (Barranquilla)*, *38*(1), 240–253. doi:10.14482un.38.1.614.4

Grant, H. (2015). The "Social" of Crime Prevention: Meaning and Implications of Social Crime Prevention for Police in the Developing World. In *Social Crime Prevention in the Developing World* (pp. 1–7). Springer. doi:10.1007/978-3-319-13027-9_1

Greca, I., Ortiz-Revilla, J. & Arriassecq, I. (2021). Diseño y evaluación de una secuencia de enseñanza-aprendizaje STEAM para Educación Primaria. *Revista Eureka sobre Enseñanza y Divulgación de las Ciencias, 18*(1), 1-19. doi:10.25267/Rev_Eureka_ensen_divulg_cienc.2021.v18.i1.1802

Guerrero, D. S. (2022). El debate en torno a la noción instrumental de tecnología. *Revista Ethika*, (6), 149–168. doi:10.5354/2452-6037.2022.66375

Haczek, Á. R. (2022, September 1). *Estos son los estados de EE.UU. con más y menos delitos violentos.* CNN. https://cnnespanol.cnn.com/2022/09/01/delitos-violentos-estados-unidos-orix/#:~:text=En%20Estados%20Unidos%2C%20la%20media

Halder, A., & Kumar, A. (2019). Active learning using rough fuzzy classifier for cancer prediction from microarray gene expression data. *Journal of Biomedical Informatics, 92*, 1–13. doi:10.1016/j.jbi.2019.103136 PMID:30802546

Hang, N. T., Huy, D. T. N., Le, T. H., Gwoździewicz, S., Thanh, N. T. P., Dung, N. T., & Hien, D. T. N. (2022). Further Analysis on Internet of Things (IOT) Applications in Emerging Markets and Vietnam. In Ambient Communications and Computer Systems. *Proceedings of RACCCS, 2021*, 407–416.

Haridas, A., Samad, A. A., Vysakh, D., & Pathari, V. (2022, March). A blockchain-based platform for smart contracts and intellectual property protection for the additive manufacturing industry. In *2022 IEEE International Conference on Signal Processing, Informatics, Communication and Energy Systems (SPICES)* (Vol. 1, pp. 223-230). IEEE. 10.1109/SPICES52834.2022.9774219

Hattingh, M. J. (2015). The use of Facebook by a Community Policing Forum to combat crime. *Proceedings of the 2015 Annual Research Conference on South African Institute of Computer Scientists and Information Technologists.* 10.1145/2815782.2815811

Helo, P., & Hao, Y. (2019). Blockchains in operations and supply chains: A model and reference implementation. *Computers & Industrial Engineering, 136*, 242–251. doi:10.1016/j.cie.2019.07.023

Hercheui, M., & Ranjith, R. (2020). Improving organization dynamic capabilities using artificial intelligence. *The Global Journal of Business Research, 14*(1), 87–96.

Hernández B. H. (2021). *Homicidios en América Latina y el Caribe: magnitud y factores asociados.* Academic Press.

Heuvel, S. (2014). *The demand for short-term, safe assets and financial stability: Some evidence and implications for central bank policies.* Academic Press.

Hoda, D. S., & Raich, V. (2017). *Fuzzy Logic Models and Fuzzy Control.* Alpha Science International LTD.

Ho, L. T., Gan, C., Jin, S., & Le, B. (2022). Artificial intelligence and firm performance: Does machine intelligence shield firms from risks? *Journal of Risk and Financial Management*, *15*(7), 302. doi:10.3390/jrfm15070302

Hollebeek, L., Jaakkola, E., & Alexander, M. (2018). Beyond the dyadic: Customer engagement in increasingly networked environments. *Journal of Service Management*, *29*(3), 330–332. doi:10.1108/JOSM-05-2018-410

Hong, J. I., & Lee, Y. G. (2020). Dynamic capabilities, environmental dynamism, and innovation performance: Evidence from the biotech industry. *Sustainability*, *12*(6), 2217.

Huamani, B. R. P. (2019). *It manufactures a CNC machine to make figures in low relief. National University of San Agustin de Arequipa. Faculty of Process Engineering. Academic Department of Metallurgy Engineering and Environmental Engineering*. School of Metallurgical Engineering.

Huang, H., Jedynak, B. M., & Bader, J. S. (2007). Where have all the interactions gone? Estimating the coverage of two-hybrid protein interaction maps. *PLoS Computational Biology*, *3*(11), e214. Advance online publication. doi:10.1371/journal.pcbi.0030214 PMID:18039026

Huang, W., & Hayat, A. (2019). Impact of artificial intelligence in enterprises HR performance in Pakistan: A comparison study with Australia. *Global Journal of Management and Business Research*, *19*(A15), 53–58.

Huang, Z., & Lin, D. K. (2009). The time-series link prediction problem with applications in communication surveillance. *INFORMS Journal on Computing*, *21*(2), 286–303. doi:10.1287/ijoc.1080.0292

Huffman, D., Thomas, K., & Basham, J. (2020). The Transdisciplinary Nature of STEAM Education: Integrating STEAM in Pre-Service Teacher Education. In K. Thomas & D. Huffman (Eds.), *Challenges and Opportunities for Transforming From STEM to STEAM Education* (pp. 221–237). IGI Global. doi:10.4018/978-1-7998-2517-3.ch009

Hurtado, J. C. (2015). *Design and construction of a CNC-milling machine prototype*. Carlos III University of Madrid.

Ibarra-Esquer, J. E., González-Navarro, F. F., Flores-Rios, B. L., Burtseva, L., & Astorga-Vargas, M. A. (2017). Tracking the evolution of the Internet of things concept across different application domains. *Sensors (Basel)*, *17*(6), 1379. doi:10.339017061379 PMID:28613238

IEEE Standards Association. (2023, February 14). *Internet of things*. IEEE Standards Association. Retrieved March 23, 2023, from https://standards.ieee.org/initiatives/iot/

Iglesias, Q. F., Aguilera, L. T., Padilla, B. A., Madaf, P. M., & Ten, C. E. (2022). *Optical evaluation of surface roughness in wood parts processed by robotic sanding*. Department of Mechanical Engineering, Faculty of Engineering and Sciences, University of La Frontera, Temuco, Chile and Forest Products Laboratory, Institute of Forests and Society, Faculty of Forest Sciences and Natural Resources, Universidad Austral de Chile, Valdivia, Chile.

INEI. (2016). *Instituto Nacional de Estadistica e Informatica.* https://m.inei.gob.pe/prensa/noticias/en-el-ano-2016-la-tasa-de-homicidios-fue-de-77-muertes-por-cada-100-mil-habitantes-9856/

Iqbal, M. J., & Khizar, H. M. U. (2022). The epistemology of entrepreneurial responsible orientation (ERO): Theory, conceptualization, and future research. *SAGE Open*, *12*(2). doi:10.1177/21582440221089320

Ivars, P., Fernández, C., & Llinares, S. (2020). Uso de una trayectoria hipotética de aprendizaje para proponer actividades de instrucción. *Enseñanza de las Ciencias*, *38*(3), 105–124. doi:10.5565/rev/ensciencias.2947

Jackson, H., Basham, J., Thomas, K., & Hunt, C. (2020). Designing STEAM Learning Environments. In K. Thomas & D. Huffman (Eds.), *Challenges and Opportunities for Transforming From STEM to STEAM Education* (pp. 1–23). IGI Global. doi:10.4018/978-1-7998-2517-3.ch001

Jensen, E., & Hanghøj, T. (2020). What's the math in Minecraft? A Design-Based Study of Students' Perspectives and Mathematical Experiences Across game and School Domains. *Electronic Journal of e-Learning*, *18*(3), 261–274. doi:10.34190/EJEL.20.18.3.005

Jiang, Z., Yuan, S., Ma, J., & Wang, Q. (2022). The evolution of production scheduling from Industry 3.0 through Industry 4.0. *International Journal of Production Research*, *60*(11), 3534–3554. doi:10.1080/00207543.2021.1925772

Juškevičienė, A., Dagienė, V., & Dolgopolovas, V. (2021). Integrated activities in STEM environment: Methodology and implementation practice. *Computer Applications in Engineering Education*, *29*(1), 209–228. doi:10.1002/cae.22324

Karpunina, E. K., Yurina, E. A., Cheremisina, T. N., Andryiashka, M. V., & Kuzmina, O. Y. (2019). 'Conversation Economy' as a determinant of digital development: New opportunities and latent threats. *Proceedings of the 34th International Business Information Management Association Conference.*

Kashif, M., Arshad, S., Tahir, M., Umair, M., & Waqas, P. (2018). A Systematic Review of Cyber Security and Classification of Attacks in Networks. International Journal of Advanced Computer Science and Applications, 9(6). doi:10.14569/IJACSA.2018.090629

Kassim, M. R. M. (2020, November). IoT applications in smart agriculture: Issues and challenges. In *2020 IEEE conference on open systems (ICOS)* (pp. 19-24). IEEE. doi:10.1109/ICOS50156.2020.9293672

Keefe, E. M., Thomas, J. A., Buller, G. A., & Banks, C. E. (2022). Textile additive manufacturing: An overview. *Cogent Engineering*, *9*(1), 2048439. doi:10.1080/23311916.2022.2048439

Khalid, S., Khan, S. A., & Ifzal, S. Q. (2021). A Fuzzy Logic-Based Framework for Mapping Crime Data on Established Sociological Hypothesis for Societal Disorder Identification and Prevention. *IEEE Access : Practical Innovations, Open Solutions*, *9*, 80197–80207. doi:10.1109/ACCESS.2021.3083542

Khan, A. W., Zaib, S., Khan, F., Tarimer, I., Seo, J. T., & Shin, J. (2022). Analyzing and Evaluating Critical Cyber Security Challenges Faced by Vendor Organizations in Software Development: SLR Based Approach. In IEEE Access (Vol. 10, pp. 65044–65054). Institute of Electrical and Electronics Engineers (IEEE). doi:10.1109/ACCESS.2022.3179822

Khan, A., Al-Zahrani, A., Al-Harbi, S., Al-Nashri, S., & Khan, I. A. (2018, February). Design of an IoT smart home system. In *2018 15th Learning and Technology Conference (L&T)* (pp. 1–5). IEEE. 10.1109/LT.2018.8368484

Klahn, C., Fontana, F., Leutenecker-Twelsiek, B., & Meboldt, M. (2020). Mapping value clusters of additive manufacturing on design strategies to support part identification and selection. *Rapid Prototyping Journal*, *26*(10), 1797–1807. doi:10.1108/RPJ-10-2019-0272

Kollu, P. K., Kumar, K., Kshirsagar, P. R., Islam, S., Naveed, Q. N., Hussain, M. R., & Sundramurthy, V. P. (2022). Development of advanced artificial intelligence and IoT automation in the crisis of COVID-19 Detection. *Journal of Healthcare Engineering*.

Kosmatos, E. A., Tselikas, N. D., & Boucouvalas, A. C. (2011). Integrating RFIDs and Smart Objects into a UnifiedInternet of Things Architecture. *Advances in Internet of Things*, *01*(01), 5–12. doi:10.4236/ait.2011.11002

Krčo, S., Pokrić, B., & Carrez, F. (2014, March). Designing IoT architecture (s): A European perspective. In *2014 IEEE World Forum on Internet of Things (WF-IoT)* (pp. 79-84). IEEE. doi:10.1109/WF-IoT.2014.6803124

Krüger, M., & Chiappe, A. (2021). 21st-century skills and their relationshio to STEAM learning environments: A review. [RED]. *Revista de Educación a Distancia*, *21*(68). doi:10.6018/red.470461

Krundyshev, V., & Kalinin, M. (2020). Prevention of cyber attacks in smart manufacturing applying modern neural network methods. *IOP Conference Series. Materials Science and Engineering*, *940*(1), 012011. doi:10.1088/1757-899X/940/1/012011

Kshetri, N. (2018). 1 blcokchain's roles in meeting key supply chain management objectives. *International Journal of Information Management*, *39*, 80–89. doi:10.1016/j.ijinfomgt.2017.12.005

Kulkarni, P., Kumar, A., Chate, G., & Dandannavar, P. (2021). Elements of additive manufacturing technology adoption in small- and medium-sized companies. *Innovation & Management Review*, *18*(4), 400–416. doi:10.1108/INMR-02-2020-0015

Kumar, N. (2019). Fog computing: Issues and challenges in security and privacy. *Journal of Ambient Intelligence and Humanized Computing*, *10*(3), 1183–1198.

Kurpjuweit, S., Schmidt, C. G., Klöckner, M., & Wagner, S. M. (2021). Blockchain in additive manufacturing and its impact on supply chains. *Journal of Business Logistics*, *42*(1), 46–70. doi:10.1111/jbl.12231

La Porte, N. (2018). We're not giving the answers so much as asking the questions. *Fast Company*, *224*, 46–50.

Laffi, M., & Boschma, R. (2022). Does a local knowledge base in Industry 3.0 foster diversification in Industry 4.0 technologies? Evidence from European regions. *Papers in Regional Science*, *101*(1), 5–35. doi:10.1111/pirs.12643

Lai, X., Wu, X., Zhang, L., Lu, W., & Zhong, C. (2019). Imputations of missing values using a tracking-removed autoencoder trained with incomplete data. *Neurocomputing*, *366*, 54–65. doi:10.1016/j.neucom.2019.07.066

Lampropoulos, G., Siakas, K., & Anastasiadis, T. (2018). Internet of Things (IoT) in Industry: Contemporary Application Domains, Innovative Technologies and Intelligent Manufacturing. *International Journal of Advances in Scientific Research and Engineering*, *4*(10), 109–118. doi:10.31695/IJASRE.2018.32910

Lawnik, M., & Kapczynski, A. (2019). The application of modified Chebyshev polynomials in asymmetric cryptography. In Computer Science (Vol. 20, Issue 3, p. 367). AGHU University of Science and Technology Press. doi:10.7494/csci.2019.20.3.3307

Lee Benson, N. (1985). *Mexico and the Spanish Cortes, 1810-1822. Eight Essays, Aus tin, University of Texas Press, 1966, que fue traducido al español: México y las Cortes españolas 1810-1822.* Ocho ensayos, México, Instituto de Investigaciones Legislativas, Cámara de Dipu tados, 1985.

Lee, H., Lee, Y., Lee, K., & Yim, K. (2016). Security Assessment on the Mouse Data using Mouse Loggers. In Advances on Broad-Band Wireless Computing, Communication and Applications (pp. 387–393). Springer International Publishing. doi:10.1007/978-3-319-49106-6_37

Lee, J., Davari, H., Singh, J., & Pandhare, V. (2018). Industrial artificial intelligence for industry 4.0-based manufacturing systems. *Manufacturing Letters*, *18*, 20–23. doi:10.1016/j.mfglet.2018.09.002

Lennvall, T., Gidlund, M., & Akerberg, J. (2017). Challenges when bringing IoT into industrial automation. 2017 IEEE AFRICON: Science. *Technology and Innovation for Africa, AFRICON, 2017*, 905–910. doi:10.1109/AFRCON.2017.8095602

Li, J. (2015). The Research and Application of Multi-Firewall Technology in Enterprise Network Security. International Journal of Security and Its Applications, 9(5), 153–162. doi:10.14257/ijsia.2015.9.5.16

Li, Y., & Chen, T. (2022). Blockchain empowers supply chains: challenges, opportunities and prospects. *Nankai Business Review International*. doi:10.1108/NBRI-06-2022-0066

Liben-Nowell, D., & Kleinberg, J. (2007). The link-prediction problem for social networks. *Journal of the American Society for Information Science and Technology*, *58*(7), 1019–1031. doi:10.1002/asi.20591

Li, H. (2021). Fog computing for industrial IoT: A review of enabling technologies and applications. *IEEE Internet of Things Journal*, *8*(9), 7426–7442.

Li, H., Yang, H., Zhou, L., & Jiang, C. (2021). Fog computing for industrial IoT: A review of enabling technologies and applications. *Journal of Ambient Intelligence and Humanized Computing*, *12*(2), 2109–2122.

Li, M. (2018). Fog computing for big data analytics: A survey. *Journal of Network and Computer Applications*, *98*, 18–28.

Liston, M., Morrin, A. M., Furlong, T., & Griffin, L. (2022). Integrating Data Science and the Internet of Things Into Science, Technology, Engineering, Arts, and Mathematics Education Through the Use of New and Emerging Technologies. In *Frontiers in Education* (p. 348). Frontiers. doi:10.3389/feduc.2022.757866

Liu, Y., & Kou, Z. (2007). Predicting who rated what in large-scale datasets. *SIGKDD Explorations*, *9*(2), 62–65. doi:10.1145/1345448.1345462

Liu, Y., Wu, Y., & Zhang, W. (2021). A smart contract-based secure and efficient data trading scheme for vehicular ad hoc networks. *Journal of Network and Computer Applications*, *180*, 102934. doi:10.1016/j.jnca.2021.102934

Li, W., Andreina, S., Bohli, J.-M., & Karame, G. (2017). Securing Proof-of-Stake Blockchain Protocols. In J. Garcia-Alfaro, G. Navarro-Arribas, H. Hartenstein, & J. Herrera-Joancomartí (Eds.), *Data Privacy Management, Cryptocurrencies and Blockchain Technology* (pp. 297–315). Lecture Notes in Computer Science. Springer International Publishing. doi:10.1007/978-3-319-67816-0_17

Li, X., Jiang, P., Chen, T., Luo, X., & Wen, Q. (2017). A survey on the security of blockchain systems. *Future Generation Computer Systems.*

Li, X., Li, Y., Li, X., & Li, C. (2021). A novel smart contract-based secure data sharing framework for industrial internet of things. *Journal of Industrial Information Integration*, *21*, 100177. doi:10.1016/j.jii.2021.100177

Li, Y., & Liu, Q. (2021). A comprehensive review study of cyber-attacks and cyber security; Emerging trends and recent developments. In *Energy Reports* (Vol. 7, pp. 8176–8186). Elsevier BV. doi:10.1016/j.egyr.2021.08.126

Li, Y., Li, X., Li, C., & Li, X. (2021). A smart contract-based secure data sharing scheme for cloud-assisted healthcare systems. *Journal of Medical Systems*, *45*(2), 22. doi:10.100710916-018-1142-z PMID:33426569

Li, Y., Wang, K., Xiao, Y., & Froyd, J. E. (2020). Research and trends in STEM education: A systematic review of journal publications. *International Journal of STEM Education*, *7*(1), 11. doi:10.118640594-020-00207-6

Li, Y., Xiao, Y., Wang, K., Zhang, N., Pang, Y., Wang, R., Qi, C., Yuan, Z., Xu, J., Nite, S. B., & Star, J. R. (2022). A systematic review of high impact empirical studies in STEM education. *International Journal of STEM Education*, *9*(1), 72. doi:10.118640594-022-00389-1

Lopez Taborda, L. L., Maury, H., & Pacheco, J. (2021). Design for additive manufacturing: A comprehensive review of the tendencies and limitations of methodologies. *Rapid Prototyping Journal*, *27*(5), 918–966. doi:10.1108/RPJ-11-2019-0296

Lopez-Martin, M., Carro, B., Sanchez-Esguevillas, A., & Lloret, J. (2017). Network traffic classifier with convolutional and recurrent neural networks for Internet of Things. *IEEE Access : Practical Innovations, Open Solutions*, *5*, 18042–18050. doi:10.1109/ACCESS.2017.2747560

López-Pérez, E. (2020). *La industria 4.0 y las nuevas formas de trabajar: una perspectiva desde el caso mexicano en tiempos del COVID 19*. Academic Press.

Lorenzo-Lledó, A. (2023). El modelo educativo inclusivo desde el Diseño Universal para el Aprendizaje (DUA). *Atención a la Diversidad en los Centros Educativos*. [Online]. Disponible en: https://rua.ua.es/dspace/bitstream/10045/131375/1/El-modelo-educativo-inclusivo-desde-el-Diseno-Universal-para-el-Aprend
izaje-DUA.pdf

Lu, Y. (2018). Cybersecurity Research: A Review of Current Research Topics. Journal of Industrial Integration and Management, 3(4). doi:10.1142/S2424862218500148

Lutz, Roth, Sefi, Bondy, & Malvina. (2019). Crime is in the air: the contemporaneous relationship between air pollution and crime. Harvard Dataverse. doi:10.7910/DVN/RMMITA

MacDonald, A., Wise, K., Tregloan, K., Fountain, W., Wallis, L., & Holmstrom, N. (2020). Designing STEAM education: Fostering relationality through design-led disruption. *International Journal of Art & Design Education*, *39*(1), 227–241. doi:10.1111/jade.12258

Madakam, S., Lake, V., Lake, V., & Lake, V. (2015). Internet of Things (IoT): A literature review. *Journal of Computer and Communications*, *3*(05), 164–173. doi:10.4236/jcc.2015.35021

Maglaras, L., Janicke, H., & Ferrag, M. A. (2022). Cybersecurity of Critical Infrastructures: Challenges and Solutions. In Sensors (Vol. 22, Issue 14, p. 5105). MDPI AG. doi:10.339022145105

Margot, K., & Kettler, T. (2019). Teachers' perception of STEM integration and education: A systematic literature review. *International Journal of STEM Education*, *6*(1), 2. doi:10.118640594-018-0151-2

Martínez, P. (2006). El método de estudio de caso: estrategia metodológica de la investigación científica. *Pensamiento & gestión*, (20), 165-193. https://www.redalyc.org/pdf/646/64602005.pdf

Martínez, S. T. (2000). *The Tlaxcalan Expansion to the New Spanish Colonial North: The Stages of Agricultural Expansion*. Colegio de Postgraduados. https://www.researchgate.net/publication/269410739_la_Diaspo ra_Tlaxcalteca

Mathew, A. R. (2019). Cyber Security through Blockchain Technology. International Journal of Engineering and Advanced Technology, 9(1), 3821–3824. doi:10.35940/ijeat.A9836.109119

Mathur, A., Dabas, A., & Sharma, N. (2022). Evolution From Industry 1.0 to Industry 5.0. In *2022 4th International Conference on Advances in Computing, Communication Control and Networking (ICAC3N)* (pp. 1390-1394). IEEE.

Mayo Clinic. (n.d.), *Dissociative disorders*. Dissociative Disorders. https://www.mayoclinic.org/diseases-conditions/dissociative-disorders/symptoms-causes/syc-20355215

McCracken, H. (2017). The great AI war of 2018. *Fast Company*, *220*, 64–73.

McLure, F., Tang, K., & Williams, P. (2022). What do integrated STEM projects look like in middle school and high school classrooms? A systematic literature review of empirical studies of iSTEM projects. *International Journal of STEM Education*, *9*(1), 73. doi:10.118640594-022-00390-8

Medina-Valdés, Z., Díaz-Roque, J., & Plaza-Macías, N. M. (2022). La Cuarta Revolución Industrial y el sistema de contradicciones del capitalismo. *Economia y Desarrollo*, *166*(1).

Mehta, R., Singh, H., Banerjee, A., Bozhuk, S., & Kozlova, N. (2020). Comparative analysis of the consequences of purchasing models transformation within the global digitalization of the economy. *IOP Conference Series. Materials Science and Engineering*, *940*(1), 012071. doi:10.1088/1757-899X/940/1/012071

Mellado, D., Mouratidis, H., & Fernández-Medina, E. (2014). Secure Tropos framework for software product lines requirements engineering. In Computer Standards & Interfaces (Vol. 36, Issue 4, pp. 711–722). Elsevier BV. doi:10.1016/j.csi.2013.12.006

Meneghello, F., Calore, M., Zucchetto, D., Polese, M., & Zanella, A. (2019). IoT: Internet of Threats? A Survey of Practical Security Vulnerabilities in Real IoT Devices. *IEEE Internet of Things Journal*, *6*(5), 8182–8201. doi:10.1109/JIOT.2019.2935189

Mikalef, P., Fjørtoft, S. O., & Torvatn, H. Y. (2019). Developing an artificial intelligence capability: A theoretical framework for business value. Business Information Systems Workshops: BIS 2019 International Workshops. doi:10.1007/978-3-030-36691-9_34

Miller, T. (2018). Developing numeracy skills using interactive technology in a play-based learning environment. *International Journal of STEM Education*, *5*(1), 39. doi:10.118640594-018-0135-2 PMID:30631729

Mirshahvalad, A., Lindholm, J., Derlén, M., & Rosvall, M. (2012). Significant communities in large sparse networks. *PLoS One*, *7*(3), e33721. Advance online publication. doi:10.1371/journal.pone.0033721 PMID:22479433

Mkrttchian, V., Gamidullaeva, L., & Galiautdinov, R. (2019). *Design of Nano-scale Electrodes and Development of Avatar-Based Control System for Energy-Efficient Power Engineering: Application of an Internet of Things and People (IOTAP) Research Center. International Journal of Applied Nanotechnology Research.* doi:10.4018/IJANR.2019010104

Mohamed, K. F., & Azer, M. A. (2022). Malware Detection Techniques. In *2022 4th Novel Intelligent and Leading Emerging Sciences Conference (NILES). 2022 4th Novel Intelligent and Leading Emerging Sciences Conference (NILES).* IEEE. 10.1109/NILES56402.2022.9942395

Mohsin, M., Anwar, Z., Zaman, F., & Al-Shaer, E. (2017). IoTChecker: A data-driven framework for security analytics of Internet of Things configurations. In Computers & Security (Vol. 70, pp. 199–223). Elsevier BV. doi:10.1016/j.cose.2017.05.012

Molina P. A. F. (2012). *Design, construction and programming of a prototype of a CNC machine for milling and punching PCB printed circuit boards.* University of Nariño. Faculty of Engineering. Electronic Engineering Program.

Mondal, D. (2019). The Internet of Things (IoT) and Industrial Automation: A future perspective. *World Journal of Modelling and Simulation, 15*(2), 140–149. Retrieved from www.rcciit.org doi:10.1016/j.cirp.2016.06.005

Mori, J., Kajikawa, Y., Kashima, H., & Sakata, I. (2012). Machine learning approach for finding business partners and building reciprocal relationships. *Expert Systems with Applications, 39*(12), 10402–10407. doi:10.1016/j.eswa.2012.01.202

Muñoz, J, C. (2022). *Sobre la enseñabilidad de la administración. Algunas reflexiones preliminares.* Academic Press.

Nafea, R. A., & Amin Almaiah, M. (2021). Cyber Security Threats in Cloud: Literature Review. In *2021 International Conference on Information Technology (ICIT). 2021 International Conference on Information Technology (ICIT).* IEEE. 10.1109/ICIT52682.2021.9491638

Nagy, J., Oláh, J., Erdei, E., Máté, D., & Popp, J. (2018). The role and impact of industry 4.0 and the internet of things on the business strategy of the value chain-the case of Hungary. *Sustainability (Basel), 10*(10), 3491. Advance online publication. doi:10.3390u10103491

Naseer, M., Rusdi, J. F., Shanono, N. M., Salam, S., Muslim, Z. B., Abu, N. A., & Abadi, I. (2021). Malware Detection: Issues and Challenges. Journal of Physics: Conference Series, 1807(1). doi:10.1088/1742-6596/1807/1/012011

Nasir, M., Summerfield, N., Dağ, A., & Öztekin, A. (2020). A service analytic approach to studying patient no-shows. *Service Business, 14*(2), 287–313. doi:10.100711628-020-00415-8

Nasri, N., Rahimi, N., Nasri, N. & Talib, M. (2021). A Comparison Study between Universal Design for Learning-Multiple Intelligence (UDL-MI) Oriented STEM Program and Traditional STEM Program for Inclusive Education. *Sustainability, 13*(2), 554. MDPI AG. doi:10.3390/su13020554

National Geographic. (2022). *Map | National Geographic Society*. Education.nationalgeographic.org. https://education.nationalgeographic.org/resource/map

NCI. (2022). *National Cancer Institute, Adult Non-Hodgkin Lymphoma Treatment (PDQ®)–Patient Version*. NCI. https://www.cancer.gov/types/lymphoma/patient/adult-nhl-treatment-pdq,

Nguyen, T. T., Nguyen, Q. V. H., Nguyen, D. T., Yang, S., Eklund, P. W., Huynh-The, T., Nguyen, T. T., Pham, Q. V., Razzak, I., & Hsu, E. B. (2020). Artificial intelligence in the battle against coronavirus (COVID-19): A survey and future research directions. arXiv preprint arXiv:2008.07343

Nguyen, C. T., Hoang, D. T., Nguyen, D. N., Niyato, D., Nguyen, H. T., & Dutkiewicz, E. (2019). Proof-of-Stake Consensus Mechanisms for Future Blockchain Networks: Fundamentals, Applications and Opportunities. *IEEE Access : Practical Innovations, Open Solutions*, *7*, 85727–85745. doi:10.1109/ACCESS.2019.2925010

Niesen, T., Houy, C., Fettke, P., & Loos, P. (2016, January). Towards an integrative big data analysis framework for data-driven risk management in industry 4.0. In *2016 49th Hawaii international conference on system sciences (HICSS)* (pp. 5065–5074). IEEE. 10.1109/HICSS.2016.627

NLM. (2022). *Capgras Syndrome*. National Library of Medicine, National Center for Biotechnology Information. https://pubmed.ncbi.nlm.nih.gov/34033319/

Nolan, A. (2020). Artificial intelligence, digital technology and advanced production. In OECD (Ed.), The Digitalisation of Science, Technology and Innovation: Key Developments and Policies (pp. 119–142). OECD Publishing.

Nowotarski, P., & Paslawski, J. (2017, October). Industry 4.0 concept introduction into construction SMEs. In IOP Conference Series: Materials Science and Engineering (Vol. 245, No. 5, p. 052043). IOP Publishing. doi:10.1088/1757-899X/245/5/052043

Öberg, C. (2019). Additive manufacturing – digitally changing the global business landscape. *European Journal of Management and Business Economics*, *28*(2), 174–188. doi:10.1108/EJMBE-11-2018-0116

Oh, J., & Shong, I. (2017). A case study on business model innovations using Blockchain: Focusing on financial institutions. *Asia Pacific Journal of Innovation and Entrepreneurship*, *11*(3), 335–344. doi:10.1108/APJIE-12-2017-038

Ohnishi, Y., Honda, K., Nishioka, R., Mori, S. & Kawada, K. (2017). Robotics programming learning for elementary and junior high school students. *Journal of robotics and mechatronics, 29*(6), 992-998. doi:10.20965/jrm.2017.p0992

Okano, M. T. (2017). IOT and Industry 4.0: The Industrial New Revolution. *ICMIS-17 - International Conference on Management and Information Systems*, 75–82.

Okul, Ş., Aksu, D., & Orman, Z. (2019). Investigation of artificial intelligence based optimization algorithms. *Journal of Istanbul Sabahattin Zaim University Natural Sciences Institute*, *1*(1), 11–16.

ONU-Habitat. (2022). *Reducción del crimen a través de la planificación y gestión urbana.* https://onuhabitat.org.mx/index.php/reduccion-del-crimen-a-traves-de-la-planificacion-y-gestion-urbana

Oudani, M., Sebbar, A., Zkik, K., El Harraki, I., & Belhadi, A. (2022). Green block-chain based IoT for secured supply chain of hazardous materials. *Computers & Industrial Engineering, 108814.* Advance online publication. doi:10.1016/j.cie.2022.108814

Palmer, D., Warren, I., & Miller, P. (2018). Crime prevention, dataveillance, and the regulation of information communication technologies. In *Encyclopedia of Social Network Analysis and Mining* (pp. 463–474). Springer New York. doi:10.1007/978-1-4939-7131-2_210

Papadopoulos, T., Singh, S. P., Spanaki, K., Gunasekaran, A., & Dubey, R. (2022). Towards the next generation of manufacturing: Implications of big data and digitalization in the context of industry 4.0. *Production Planning and Control, 33*(2-3), 101–104. doi:10.1080/09537287.2020.1810767

Parida, V., & Madsen, E. L. (2020). Artificial intelligence for dynamic capabilities: Empirical evidence from manufacturing firms. International Journal of Operations & Production Management.

Park, S. (2016). Development of innovative strategies for the Korean manufacturing industry by use of the Connected Smart Factory (CSF). *Procedia Computer Science, 91*, 744–750. doi:10.1016/j.procs.2016.07.067

Parvanda, R., & Kala, P. (2022). Trends, opportunities, and challenges in the integration of the additive manufacturing with Industry 4.0. *Progress in Additive Manufacturing*, 1-28. doi:10.1007/s40964-022-00351-1

Pearce, S. (2021, June). *Smart contracts: The good, the bad and the ugly.* TechRadar. https://www.techradar.com/news/smart-contracts-the-good-the-bad-and-the-ugly

Pereira, A. C., & Romero, F. (2017). A review of the meanings and the implications of the Industry 4.0 concept. *Procedia Manufacturing, 13*, 1206–1214. doi:10.1016/j.promfg.2017.09.032

Pereira, F., Correia, R., Pinho, P., Lopes, S. I., & Carvalho, N. B. (2020). Challenges in resource-constrained IoT devices: Energy and communication as critical success factors for future IoT deployment. *Sensors (Basel), 20*(22), 6420. doi:10.339020226420 PMID:33182749

Peterson, B., & Hipple, B. (2020). Formative Assessment in Hands-On STEM Education. In C. Martin, D. Polly, & R. Lambert (Eds.), *Handbook of Research on Formative Assessment in Pre-K Through Elementary Classrooms* (pp. 1–29). IGI Global. doi:10.4018/978-1-7998-0323-2.ch009

Prakash, R., Anoop, V. S., & Asharaf, S. (2022). Blockchain technology for cybersecurity: A text mining literature analysis. *International Journal of Information Management Data Insights, 2*(2), 100112. doi:10.1016/j.jjimei.2022.100112

Purwania, I. B. G., Kumara, I. N. S., & Sudarma, M. (2020). Application of IoT-Based System for Monitoring Energy Consumption. *International Journal of Engineering and Emerging Technology*, *5*(2), 81–93.

Quigley, C., Herro, D., & Jamil, F. (2017). Developing a conceptual model of STEAM teaching practices. *School Science and Mathematics*, *117*(1-2), 1–12. doi:10.1111sm.12201

Rajasekharaiah, K. M., Dule, C. S., & Sudarshan, E. (2020). Cyber Security Challenges and its Emerging Trends on Latest Technologies. In IOP Conference Series: Materials Science and Engineering (Vol. 981, Issue 2, p. 022062). IOP Publishing. doi:10.1088/1757-899X/981/2/022062

Rajkumar, R. R., Lee, I., Sha, L., & Stankovic, J. (2010). Proceedings of the 47th design automation conference. *Cyber-physical systems: The next computing revolution*. 10.1145/1837274.1837461

Ramachandran, R., Babu, V., & Murugesan, V. P. (2022). The role of blockchain technology in the process of decision-making in human resource management: a review and future research agenda. Business Process Management Journal. doi:10.1108/BPMJ-07-2022-0351

Ramesh, P., & Vinodh, S. (2021). Analysis of factors influencing energy consumption of material extrusion-based additive manufacturing using interpretive structural modelling. *Rapid Prototyping Journal*, *27*(7), 1363–1377. doi:10.1108/RPJ-05-2020-0100

Ramirez, R., & Choucri, N. (2016). Improving Interdisciplinary Communication With Standardized Cyber Security Terminology: A Literature Review. In IEEE Access (Vol. 4, pp. 2216–2243). Institute of Electrical and Electronics Engineers (IEEE). doi:10.1109/ACCESS.2016.2544381

Ramos-Soto, A., & Pereira-Fariña, M. (2017). On modeling vagueness and uncertainty in data-to-text systems through fuzzy sets. arXiv preprint arXiv:1710.10093.

Rao, T., Gaddam, A., Kurni, M., & Saritha, K. (2021). Reliance on artificial intelligence, machine learning and deep learning in the Era of Industry 4.0. In Smart Healthcare System Design: Security and Privacy Aspects (pp. 281–299). Wiley.

Raza, S. (2021). Fog computing for healthcare: A review of applications, challenges, and future directions. *IEEE Journal of Biomedical and Health Informatics*, *25*(6), 1876–1893.

Raza, S., Shafi, S., Saeed, U., & Imran, M. (2021). Fog computing for healthcare: A review of applications, challenges, and future directions. *IEEE Access : Practical Innovations, Open Solutions*, *9*, 52363–52381.

Rehmani, M. H. (2017). Fog computing for smart cities: A review of challenges and solutions. *Journal of Ambient Intelligence and Humanized Computing*, *8*(1), 127–142.

Ren, Y., Chen, H., & Lu, Y. (2020). A smart contract-based approach for secure data sharing in vehicular ad hoc networks. *Journal of Parallel and Distributed Computing*, *135*, 32–40. doi:10.1016/j.jpdc.2019.08.014

Reynaldos-Grandón, K. (2020). Cambio y Liderazgo: Aprendizaje organizacional en tiempos de pandemia. *Revista Medica de Chile*, *148*(9), 1375–1377. doi:10.4067/S0034-98872020000901375 PMID:33399719

Righetti, F., Vallati, C., & Anastasi, G. (2018, June). IoT applications in smart cities: A perspective into social and ethical issues. In *2018 IEEE International Conference on Smart Computing (SMARTCOMP)* (pp. 387-392). IEEE. 10.1109/SMARTCOMP.2018.00034

Roses. S. (2019). *Logistics. Solgistika starts operations in Queretaro*. somosindustria.com

Rothbaum, B. H., Hodges, L. F., Ready, D., Graap, K., & Alarcon, R. D. (2001). Virtual reality exposure therapy for Vietnam veterans with posttraumatic stress disorder. *The Journal of Clinical Psychiatry*, *62*(8), 617–622. doi:10.4088/JCP.v62n0808 PMID:11561934

S.A.S., E. L. R. (n.d.). *Los países con los índices más altos de criminalidad organizada en todo el mundo*. Diario La República. https://www.larepublica.co/globoeconomia/los-paises-con-los-indices-mas-altos-de-criminalidad-organizada-en-todo-el-mund
o-3351148

Saberi, S., Kouhizadeh, M., Sarkis, J., & Shen, L. (2019). Blockchain technology and its relationships to sustainable supply chain management. *International Journal of Production Research*, *57*(7), 2117–2135. doi:10.1080/00207543.2018.1533261

Sadoughi, F., Behmanesh, A., & Sayfouri, N. (2020). Internet of things in medicine: a systematic mapping study. *Journal of Biomedical Informatics*, 103. doi:10.1016/j.jbi.2020.103383

Saleh, F. (2021). Blockchain without Waste: Proof-of-Stake. *The Review of Financial Studies*, *34*(3), 1156–1190. doi:10.1093/rfs/hhaa075

Sanchez-Medina, A. J., & C-Sánchez, E. (2020). Using machine learning and big data for efficient forecasting of hotel booking cancellations. *International Journal of Hospitality Management*, *89*, 102546. doi:10.1016/j.ijhm.2020.102546

Sánchez-Moncayo, G. K. (2021). *El impacto económico y el uso de la transición digital como innovación en el sector minorista de comercios especializados y no especializados frente a la pandemia COVID–19 en la zona centro del país* [Bachelor's thesis]. Universidad Técnica de Ambato. Facultad de Ciencias Administrativas. Carrera de Organización de Empresas.

Santos, L. N., Lopez, C. J. E. A., & López, C. J. C. (2003). Design and construction of a semi-automatic machine for sanding flat metallographic samples. University of El Salvador. Faculty of Engineering and Architecture School of Mechanical Engineering.

Satyanarayanan, M. (2017). Fog computing: Principles, architectures, and applications. In *Proceedings of the 2017 Workshop on Fog Computing and the IoT* (pp. 23-27). Academic Press.

Schuelka, M., & Thyrring, T. (2022). On the question of educational purpose: Complex educational systems analysis for inclusion. *International Journal of Inclusive Education*, *26*(5), 448–465. do i:10.1080/13603116.2019.1698062

Sekerin, V. D., Slepov, V. A., Gayduk, V. I., Bank, S. V., & Kravets, E. V. (2021). Blockchain technology development as tool for enhancing security in management and protection of intellectual property rights in additive manufacturing. *Revista Geintec-Gestao Inovacao E Tecnologias, 11*(2), 1184-1200. doi:10.47059/revistageintec.v11i2.1747

Shah, K., Patel, N., Thakkar, J., & Patel, C. (2022). Exploring applications of blockchain technology for Industry 4.0. *Materials Today: Proceedings, 62*(13), 7238–7242. doi:10.1016/j.matpr.2022.03.681

Shi, Z., Kan, C., Tian, W., & Liu, C. (2021). A Blockchain-based G-code protection approach for cyber-physical security in additive manufacturing. *Journal of Computing and Information Science in Engineering, 21*(4), 041007. Advance online publication. doi:10.1115/1.4048966

Smith-Stark, T. C. (1986). *The Mexican manual language* [Unpublished paper]. Colegio de México.

Soh, P. H., & Sia, S. K. (2019). A dynamic capabilities perspective on the strategic value of artificial intelligence: Implications for organizational learning and innovation. *The Journal of Strategic Information Systems, 28*(1), 1–12.

Soumyalatha, S. G. H. (2016, May). Study of IoT: understanding IoT architecture, applications, issues and challenges. *International Journal of Advanced Networking & Applications, 478.*

Staff, H. B. R. (2021, July). Smart Contracts: Hype vs. Reality. *Harvard Business Review*. https://hbr.org/2021/07/smart-contracts-hype-vs-reality

Stellantis. (2023). *Autonomous driving*. Stellantis. https://www.stellantis.com/en/technology/autonomous-driving

Strategic National Programs. (n.d.). *Programas Nachionales Estraegicous.* SNP. https://conacyt.mx/pronaces/

Sun, C. C., Sebastian Cardenas, D. J., Hahn, A., & Liu, C. C. (2021). Intrusion detection for cybersecurity of Smart Meters. *IEEE Transactions on Smart Grid, 12*(1), 612–622. doi:10.1109/TSG.2020.3010230

Szabo, N. (1997). *The idea of smart contracts*. Nick Szabo's Papers and Concise Tutorials.

Tabassum, F., & Ghosh, N. (2018). Future of performance management in artificial intelligence ERA. *Shanlax International Journal of Arts. Science and Humanities, 6*(S1), 7–14. doi:10.5281/zenodo.1469447

Tabassum, K., Ibrahim, A., & El Rahman, S. A. (2019, April). Security issues and challenges in IoT. In *2019 International Conference on Computer and Information Sciences (ICCIS)* (pp. 1–5). IEEE.

Tasca, P., & Tessone, C. J. (2019). A Taxonomy of Blockchain Technologies: Principles of Identification and Classification. *Ledger, 4*. doi:10.5195/ledger.2019.140

Teece, D. J. (2017). Explicating dynamic capabilities: The nature and microfoundations of (sustainable) enterprise performance. *Strategic Management Journal*, *28*(13), 1319–1350. doi:10.1002mj.640

Teece, D. J., Peteraf, M., & Leih, S. (2016). Dynamic capabilities and organizational agility: Risk, uncertainty, and strategy in the innovation economy. *California Management Review*, *58*(4), 13–35. doi:10.1525/cmr.2016.58.4.13

Teece, D. J., Pisano, G., & Shuen, A. (1997). Dynamic capabilities and strategic management. *Strategic Management Journal*, *18*(7), 509–533. doi:10.1002/(SICI)1097-0266(199708)18:7<509::AID-SMJ882>3.0.CO;2-Z

Tekerek, M., & Tekerek, B. (2018). Integrated instructional material and development processes. *Turkish Journal of Education*, *7*(3), 156–168. doi:10.19128/turje.362491

Tessaro, S. (2016). Sloth: A Lightweight Delay Function for ASIC-Resistant Proof-of-Work Based on Nakamoto's Consensus Algorithm. *Journal of Cryptology*, *29*(2), 613–638.

Thakur, K., Qiu, M., Gai, K., & Ali, M. L. (2015). An Investigation on Cyber Security Threats and Security Models. In *2015 IEEE 2nd International Conference on Cyber Security and Cloud Computing. 2015 IEEE 2nd International Conference on Cyber Security and Cloud Computing (CSCloud).* IEEE. 10.1109/CSCloud.2015.71

Thomas, D. (2016). Costs, benefits, and adoption of additive manufacturing: A supply chain perspective. *International Journal of Advanced Manufacturing Technology*, *85*(5), 1857–1876. doi:10.100700170-015-7973-6 PMID:28747809

Timothy, J. R. (2017). Fuzzy Logic with Engineering Applications. John Wiley and Sons, Ltd.

Torres, G. E. P., & Granadillo, A. (2019). *Estudio Dogmático y filosófico del concepto de delito*. J.M Bosch. doi:10.2307/j.ctvq2vx91

Troyanskaya, O., Cantor, M., Sherlock, G., Brown, P., Hastie, T., Tibshirani, R., Botstein, D., & Altman, R. B. (2001). Missing value estimation methods for DNA microarrays. *Bioinformatics (Oxford, England)*, *17*(6), 520–525. doi:10.1093/bioinformatics/17.6.520 PMID:11395428

Turkcan, H., Imamoglu, S. Z., & Ince, H. (2022). To be more innovative and more competitive in dynamic environments: The role of additive manufacturing. *International Journal of Production Economics*, *246*, 108418. doi:10.1016/j.ijpe.2022.108418

U.S. Crime Index State Rank. (n.d.). http://www.usa.com/rank/us--crime-index--state-rank.htm

Urgiles, B., Tixi, K., Allauca, M., & Fonseca, G. (2021). Systematic review of the use of STEAM in higher education. *MktDESCUBRE*, *1*(17), 85–94. doi:10.36779/mktdescubre.v1i17.580

Van Der Wees, A., Breeuwsma, J., & van Sleen, A. (2022). IoT Societal Impact–Legal Considerations and Perspectives. In Digitising the Industry Internet of Things Connecting the Physical, Digital and VirtualWorlds (pp. 215-235). River Publishers. doi:10.1201/9781003337966-7

Van-Gigch, J. P. (2002). Comparing the epistemologies of scientific disciplines in two distinct domains: modern physics versus social sciences. II: epistemology and knowledge characteristics of the 'new'social sciences. *Systems Research and Behavioral Science*, *19*(6), 551–562. doi:10.1002res.466

Veerasamy, B., & Sangeetha, C. M. (2022). MB-FL: Macro-Block Fuzzy Logic for Video Compression in Multimedia Applications. *International Journal of Fuzzy Logic and Intelligent Systems*, *22*(4), 366–372. doi:10.5391/IJFIS.2022.22.4.366

Veksler, V. D., Buchler, N., Hoffman, B. E., Cassenti, D. N., Sample, C., & Sugrim, S. (2018). Simulations in Cyber-Security: A Review of Cognitive Modeling of Network Attackers, Defenders, and Users. In Frontiers in Psychology (Vol. 9). Frontiers Media SA. https://doi.org/ doi:10.3389/fpsyg.2018.0069

Venkatram, K., & Geetha, M. A. (2017). Review on big data analytics – Concepts, philosophy, process and applications. *Cybernetics and Information Technologies*, *17*(2), 3–27. doi:10.1515/cait-2017-0013

Verma, A., & Shri, C. (2022). Cyber Security: A Review of Cyber Crimes, Security Challenges and Measures to Control. Vision: The Journal of Business Perspective. doi:10.1177/09722629221074760

Vermesan, O., EisenHauer, M., Serrano, M., Guillemin, P., Sundmaeker, H., Tragos, E. Z., ... Darmois, E. C. (2022). The next generation internet of things–hyperconnectivity and embedded intelligence at the edge. In Next Generation Internet of Things–Distributed Intelligence at the Edge and Human-Machine Interactions (pp. 19-102). River Publishers.

Viskupic, K., Earl, B., & Shadle, S. (2022). Adapting the CACAO model to support higher education STEM teaching reform. *International Journal of STEM Education*, *9*(1), 6. doi:10.118640594-021-00325-9

Vogel, P. S. (2021, May). *Why smart contracts may not be smart after all.* CIO Dive. https://www.ciodive.com/news/smart-contracts-limitations/599279/

Wamba-Taguimdje, S. L., Fossu Wamba, S., Kala Kamdjoug, J. R., & Tchatchouang Wanko, C. E. (2020). Influence of artificial intelligence (AI) on firm performance: The business value of AI-based transformation projects. *Business Process Management Journal*, *26*(7), 1893–1924. doi:10.1108/BPMJ-10-2019-0411

Wang, C., Chen, D., & Hu, Q. (2014). Fuzzy information systems and their homomorphisms. *Fuzzy Sets and Systems*, *249*, 128–138. doi:10.1016/j.fss.2014.02.009

Wang, L. H., Chen, B., Hwang, G. J., Guan, J.-Q., & Wang, Y.-Q. (2022). Effects of digital game-based STEM education on students' learning achievement: A meta-analysis. *International Journal of STEM Education*, *9*(1), 26. doi:10.118640594-022-00344-0

Wang, P., Xu, B. W., Wu, Y. R., & Zhou, X. Y. (2014). Link prediction in social networks: The state-of-the-art. *Science China. Information Sciences*, *58*(1), 1–38. doi:10.100711432-014-5237-y

Wang, S., Hu, Y., & Qi, G. (2022). Blockchain and deep learning based trust management for Internet of Vehicles. *Simulation Modelling Practice and Theory, 120*, 102627. doi:10.1016/j.simpat.2022.102627

Wang, Y. (2018). Fog computing for the Internet of Things: A survey. *IEEE Internet of Things Journal, 5*(2), 854–867.

Wang, Y., Wang, L., Rastegar-Mojarad, M., Moon, S., Shen, F., Afzal, N., Liu, S., Zeng, Y., Mehrabi, S., Sohn, S., & Liu, H. (2018). Clinical Information Extraction Applications: A literature review. *Journal of Biomedical Informatics, 77*, 34–49. doi:10.1016/j.jbi.2017.11.011 PMID:29162496

Wang, Y., Wang, L., & Wang, X. (2020). A smart contract-based approach for secure data sharing in cloud storage. *Journal of Network and Computer Applications, 168*, 102553. doi:10.1016/j.jnca.2020.102553

Wang, Y., Yu, C. Y., & Chan, H. H. (2012). Predicting construction cost and schedule success using artificial neural networks ensemble and support vector machines classification models. *International Journal of Project Management, 30*(4), 470–478. doi:10.1016/j.ijproman.2011.09.002

Warner, K. S. R., & Wäger, M. (2019). Building dynamic capabilities for digital transformation: An ongoing process of strategic renewal. *Long Range Planning, 52*(3), 326–349. doi:10.1016/j.lrp.2018.12.001

Webber, S. S., Detjen, J., MacLean, T. L., & Thomas, D. (2019). Team challenges: Is artificial intelligence the solution? *Business Horizons, 62*(6), 741–750. doi:10.1016/j.bushor.2019.07.007

Weiser, M. (1991). The Computer for the 21st Century. *Scientific American, 265*(3), 94–105. doi:10.1038cientificamerican0991-94 PMID:1675486

Wichmann, R., Eisenbart, B., & Gericke, K. (2019). The Direction of Industry: A Literature Review on Industry 4.0. *Proceedings of the Design Society: International Conference on Engineering Design, 1*(1), 2129–2138. doi:10.1017/dsi.2019.219

Wiljer, D., & Hakim, Z. (2019). Developing an artificial intelligence–enabled health care practice: Rewiring health care professions for better care. *Journal of Medical Imaging and Radiation Sciences, 50*(4, Supplement 2), 8–14. doi:10.1016/j.jmir.2019.09.010 PMID:31791914

Wirth, N. (2018). Hello marketing, what can artificial intelligence help you with. *International Journal of Market Research, 60*(5), 435–438. doi:10.1177/1470785318776841

Wohlgemuth, V. (2021). Dynamic capabilities: Celebrating the plurality of understandings of the concept. In T. Bayón, M. Eisend, J. Koch, A. Söllner, M. Vodosek, & H. T. Wagner (Eds.), *Dynamic Capabilities and Relationships: Discourses, Concepts, and Reflections* (pp. 1–28). Springer.

Wójcicki, K., Biegańska, M., Paliwoda, B., & Górna, J. (2022). Internet of Things in Industry: Research Profiling, Application, Challenges and Opportunities—A Review. *Energies, 15*(5), 1806. Advance online publication. doi:10.3390/en15051806

Xiong, Z., Wu, Q., Zhao, X., Zuo, J., Xiong, H., & Xie, B. (2020). the application of artificial intelligence in healthcare: A comprehensive review and critique. *Journal of Healthcare Engineering, 2020*, 1–21. doi:10.1155/2020/2973671

Xu, J., Li, X., & Li, Y. (2019). Smart contract-based secure data sharing scheme for cloud-assisted industrial Internet of Things. *Journal of Industrial Information Integration, 15*, 1–7. doi:10.1016/j.jii.2019.02.001

Xu, Z., & Zou, C. (2021). What can blockchain do and cannot do? *China Economic Journal, 14*(1), 4–25. doi:10.1080/17538963.2020.1748968

Yamany, B., Elsayed, M. S., Jurcut, A. D., Abdelbaki, N., & Azer, M. A. (2022). A New Scheme for Ransomware Classification and Clustering Using Static Features. In Electronics (Vol. 11, Issue 20, p. 3307). MDPI AG. doi:10.3390/electronics11203307

Yildiz Çankaya, S., & Sezen, B. (2019). Effects of green supply chain management practices on sustainability performance. *Journal of Manufacturing Technology Management, 30*(1), 98–121. doi:10.1108/JMTM-03-2018-0099

Yim, K. (2010). A New Noise Mingling Approach to Protect the Authentication Password. In *2010 International Conference on Complex, Intelligent and Software Intensive Systems. 2010 International Conference on Complex, Intelligent and Software Intensive Systems (CISIS)*. IEEE. 10.1109/CISIS.2010.185

Yin, B., & Yuan, C. H. (2021). Precision teaching and learning performance in a blended learning environment. *Frontiers in Psychology, 141*(12), 631125. doi:10.3389/fpsyg.2021.631125 PMID:33613404

Yousefpour, M. (2019). Application development in fog computing: A survey. *IEEE Access : Practical Innovations, Open Solutions, 7*, 112792–112809.

Zadeh, L. A. (1965). Fuzzy sets. *Information and Control, 8*(3), 338–353. doi:10.1016/S0019-9958(65)90241-X

Zadeh, L. A. (1977). Fuzzy sets and their application to pattern classification and clustering analysis. In *Classification and clustering* (pp. 251–299). Academic Press. doi:10.1016/B978-0-12-714250-0.50014-0

Zhang, R., & Chan, W. K. (2020). Evaluation of Energy Consumption in Block-Chains with Proof of Work and Proof of Stake. *Journal of Physics: Conference Series, 1584*(1), 012023. doi:10.1088/1742-6596/1584/1/012023

Zhang, H. (2018). Resource management in fog computing: A survey. *Journal of Network and Computer Applications, 103*, 1–14. doi:10.1016/j.jnca.2018.08.013

Zhang, X., Chen, X., & Xu, X. (2021). Smart contract-based privacy-preserving data sharing framework for blockchain. *Journal of Parallel and Distributed Computing, 150*, 58–66. doi:10.1016/j.jpdc.2020.11.012

Zhang, Y. C., & Yu, J. (2013). *A study on the fire IOT development strategy*. Procedia Engineering. doi:10.1016/j.proeng.2013.02.146

Zhang, Y., & Liu, X. (2019). Smart contract-based secure data sharing for cloud storage. *Journal of Cloud Computing (Heidelberg, Germany)*, *8*(1), 1–14. doi:10.118613677-019-0124-4

Zhang, Y., Sun, Y., Chen, J., & Liu, J. (2019). GasReducer: Detecting and Mitigating Gas-Costly Patterns in Ethereum Smart Contracts. *IEEE Transactions on Dependable and Secure Computing*, *17*(2), 405–419. doi:10.1109/TDSC.2018.2863135

Zhao, L. (2020). Fog-enabled vehicular networks: A review of architectures, protocols, and applications. *IEEE Communications Surveys and Tutorials*, *22*(1), 559–584.

Zhou, Z. (2019). A survey on fog computing: Architecture, key technologies, applications, and open issues. *IEEE Communications Surveys and Tutorials*, *21*(2), 1206–1258.

Zhu, L. (2019). Fog computing for multimedia applications: A review of issues and challenges. *IEEE Communications Surveys and Tutorials*, *21*(4), 3385–3408.

Zizic, M. C., Mladineo, M., Gjeldum, N., & Celent, L. (2022). From industry 4.0 towards industry 5.0: A review and analysis of paradigm shift for the people, organization and technology. *Energies*, *15*(14), 5221. doi:10.3390/en15145221

About the Contributors

Marisol Hernández Hernández is a Doctor of Computer Science and Electronic Systems. She has a post-doctorate in Technology and innovation in organizations based on fuzzy logic models and is part of the National System of Researchers in Mexico. She has professional experience in the area of information technology since 1994, when she served as head of the IT department. Her teaching experience is since 1997, at the Autonomous University of the State of Mexico, as a professor; she also served in that way at the Technological Institute of Higher Studies of Chalco. Since 2012, she has been a Full-Time Academic Technician at the Autonomous University of the State of Mexico, with administration activities of computer laboratories, General Coordination of Academic Tutorials and, currently, coordination of the Bachelor's Degree in Computer Engineering. She teaches at the bachelor's, master's, and doctoral levels, all in computer science. Her areas of interest are Augmented Reality, software engineering and fuzzy logic.

***

Caner Asbaş is founder and CEO of Hamle RF Elektrik ve Elektronik A.Ş. which is a company operating in defense, telecommunication, automotive and avionics areas. He received his undergraduate and MSc degrees from İhsan Doğramacı Bilkent University in Electrical and Electronics Engineering. He also has undergraduate degrees in International Relations from Anadolu University, Sociology from Anadolu University and Public Finance from Anadolu University. He is proficient in C, C++, C#, Java, MSSQL and Python programming languages. Asbaş continues his Ph.D. education in Management at Atılım University. His research focuses on social networks and implementation of artificial intelligence methods to social networks.

Luis Alfonso Bonilla is a Mexican researcher with a PhD in Socioeconomics, Statistics, and Informatics. He studied a Bachelor's degree in Administration at the Autonomous University of Chapingo and a Master's degree in Economics at the Division of Economic and Administrative Sciences at the Autonomous University of Chapingo. Currently, he is a member of the National System of Researchers and has a desirable profile. He is a full-time professor at the Higher Education Institute: Technological Superior Studies of Ixtapaluca.

Lizbeth Cobian Romero is an outstanding Mexican researcher. She holds a PhD in socioeconomics, statistics, and computer science, which has allowed her to carry out impactful research in her field of study. Throughout her career, she has worked in diverse business incubators, specializing in creating innovative business models and obtaining federal and state funding for the development of social projects. She is a member of the National System of Researchers, which endorses her trajectory and prestige in the scientific community. Currently, she works as a Full-time Professor at the Technological Higher Studies of Ixtapaluca, where she shares her knowledge and experience with her students.

Luis-Eduardo Faugier-Contreras has a Bachelor's degree in Industrial Relations and currently pursuing a Master's degree at DUMA University Center in the Philosophy program, with research interests revolving around administrative epistemology and social sciences. Particularly interested in topics related to aesthetics, arts, and artificial intelligence, focusing on analyzing the interactions between digital and tangible planes.

Mauricio Flores-Nicolás is a Computer Engineer by the Autonomous University of the State of Mexico (UAEMex); Master in Computer Science with a research line in educational technology by the UAEMex. During his studies, he developed a solid knowledge base on process modeling which includes the instruction of multi- and inter-disciplinary subjects applying Science, Technology, Engineering, Art and Mathematics (STEAM) that he would later apply in his professional career as a technology teacher at both high school and college level, leaving a significant footprint in his field. He is currently studying for a PhD in Computer Science at UAEMex and has participated in both national and international conferences. Beyond his professional life, he also has interests and passions outside the work environment. He enjoys long walks with his pets on foot or by bicycle, playing chess, practicing guitar and playing strategy, riddles or puzzles video games.

Rinat Galiautdinov is a Principal Software Developer and Software Architect having the expertise in Information Technology and Computer Science. Mr. Galiautdinov is also an expert in Banking/Financial industry as well as in Neurobiological sphere. Mr. Rinat Galiautdinov works on the number of highly important researches (including such the spheres as: Drones, Financial Systems and Private Money, Brain interface and neuro-biology) as an independent researcher and projects holder and developer.

Victor O. Kombo is a career diplomat, currently deployed as the Head of Foreign Relations at the Kenya Embassy in Ankara, Turkiye. He has vast experience in both private and public service from Kenya. He has good mastery in the fields of Security and Business Administration; that he has not only practiced but also engaged in advanced analysis of the same to facilitate informed strategic decision making. Victor is a frequent guest instructor to Government of Kenya institutions on matters concerning Administration and Security. He has also been keen on extending financial consultancy services, pro-bono, to start-up businesses among the youth in Kenya.

Magally Martínez-Reyes is a D. in Sciences (Educational Mathematics) from CINVESTAV, IPN, belongs to the National System of Researchers (SNI), has the PRODEP profile recognition of the SEP, and has also been awarded with the "Nota Laudatoria" by the UAEM. Full-time Professor at the University Center UAEM Valle de Chalco since 1998, where she was Coordinator of Research, of the Bachelor's Degree in Computer Engineering and of the Master's and Doctorate in Computer Science, and Director. She has published several specialized articles in national and international journals of recognized prestige in the area of mathematics and computer science, has graduated more than 20 graduate and undergraduate thesis students related to educational technology. She is a regular lecturer in several national and international congresses related to applied mathematics. She is editor of the journal El Cálculo y su enseñanza, responsible for the coordination of the Seminario Internacional sobre la Enseñanza del Cálculo, and president of the Encuentro Internacional sobre la Enseñanza del Cálculo, Ciencias y Matemáticas (EICAL).

Marva-Angélica Mora-Lumbreras is a Full-Time Professor at the Autonomous University of Tlaxcala, she obtained a PhD in Computer Science with Magna Cum Laude distinction from the Universidad de las Américas Puebla. She has published 9 books on topics such as Virtual Reality, Augmented Reality, 3D Sound and Algorithm, as well as more than 200 papers, indexed, national and international. She has titled more than 80 projects between theses and prototypes. She is currently the Editor in Charge of the Iztatl Computación Magazine and Leader of the Academic Group of Distributed Systems.

Manmohan Singh Rauthan is Professor in Computer Science and Engineering Department at H.N.B. Garhwal University, (A Central University) Uttarakhand, India. He completed his M.Sc. (Physics) from I.I.T., Delhi and M.Tech (Computer Science) from Roorkee University. He received a Ph.D. degree in (Information Technology) from Kumaun University, Nainital (India). His research interest includes information retrieval and cloud computing.

Abhijit Sarkar is an Assistant Professor in the Department of CSE(Cyber Security) of Haldia Institute of Technology with more than 11 years of teaching experience. He is presently pursuing research work in the domain of Steganography. His research interests include Steganography, Cryptography, Machine Learning, Deep Learning, Cloud Computing etc. He has published several research papers in different Journals, International Conference Proceedings.

Rajeshwari Sissodia is a Research Scholar in Computer Science and Engineering department at H.N.B. Garhwal University Uttrakhand, India. She completed B.Tech and M.Tech degree (Computer Science & Engineering) from J. B. Institute of Technology, Dehradun, India. Her research interest includes IoT & Cloud Computing.

Şule Erdem Tuzlukaya is Assoc. Prof. at the Faculty of Management, Department of Management, Atılım University. She completed her PhD in management and organization, from Başkent University. She teaches classes in organization theory and organizational behavior. Her research interests include social networks and interorganizational relationships.

Index

G

H

I

L

M

O

P

R

S

www.ingramcontent.com/pod-product-compliance
Lightning Source LLC
LaVergne TN
LVHW061536290926
849162LV00004B/813